WAVES

and

BEACHES

THE POWERFUL DYNAMICS OF SEA AND COAST

WILLARD BASCOM
KIM McCOY

patagonia

WAVES *and* BEACHES
THE POWERFUL DYNAMICS OF SEA AND COAST

Patagonia publishes a select list of titles on wilderness, wildlife, and outdoor sports that inspire and restore a connection to the natural world.

© 1964 Willard Bascom; additional text and illustrations © 2020 Kim McCoy.

Revised Third Edition

Hardcover ISBN 9781938340956
E-Book ISBN 9781938340963
Library of Congress Control Number 2020946844
Published by Patagonia Works

Editors – Makenna Goodman, John Dutton
Photo Editor – Jane Sievert
Art Director/Designer – Christina Speed
Figures – Willard Bascom, Kim McCoy, Christina Speed
Project Manager – Sonia Moore
Graphic Production – Rafael Dunn, Tausha Greenblott
Creative Director – Bill Boland
Director of Books – Karla Olson

Printed in Canada on 100 percent post-consumer recycled paper.

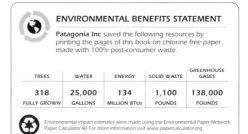

ENVIRONMENTAL BENEFITS STATEMENT

Patagonia Inc saved the following resources by printing the pages of this book on chlorine free paper made with 100% post-consumer waste.

TREES	WATER	ENERGY	SOLID WASTE	GREENHOUSE GASES
318	25,000	134	1,100	138,000
FULLY GROWN	GALLONS	MILLION BTUs	POUNDS	POUNDS

Environmental impact estimates were made using the Environmental Paper Network Paper Calculator 4.0. For more information visit www.papercalculator.org.

COVER PHOTO: *Luke Shadbolt*

FRONT END SHEET: Some waves are poorly described by words. The Pacific Ocean. *Luke Shadbolt*

FIRST SPREAD: The sea carves solid stone with powerful waves and tides. Te Whanganui-A-Hei Marine Reserve, Coromandel Peninsula, New Zealand. *Luke Shadbolt*

TITLE PHOTO: Cyclone Pam kicks up some swell on the Coromandel Peninsula, New Zealand. *Rambo Estrada*

1%
FOR THE PLANET
MEMBER

Dedication

For Anitra, Rodney, Sarah, Austin, Madelyn, Nico,

The Maltese Islands,

and

All those who love the sea.

Contents

Preface to the Third Edition *8*

Prologue *14*

1 Genesis of Land, Water, and Waves *20*

2 Ideal Waves *50*

3 Wind Waves *66*

4 Waves in Shallow Water *98*

5 Winds and Waves of Climate Change *120*

6 Tides and Seiches *132*

7 Impulsively Generated Waves *158*

8 Measuring and Making Waves *188*

9 The Surf *212*

10 Beaches: Where the Surf Meets the Sediment *248*

11 The Conveyor Belts of Sand *280*

12 Man Against the Sea *322*

Epilogue *364*

Appendices *368*

Further Reading *373*

Endnotes *376*

Glossary *388*

Acknowledgments *392*

Index *394*

A black-sand beach is the result of waves transforming hard volcanic rock into soft, black sand—one grain at a time. Iceland. *Adela Jezkova/ EyeEm via Getty Images*

Preface to the Third Edition

Generations of scientists, surfers, and sailors have used Willard Bascom's classic *Waves and Beaches*. This third edition celebrates the relationship of the sea and the land in the twenty-first century. And although time has passed since the first publication, the crashing of waves and the formation of beaches have not changed. The Sun's heat still creates the winds, which transform into waves until dying in turbulence upon a shore. However, our measurement techniques, instrumentation, and interactions with the waves have all changed. The numbers of surfers, divers, kayakers, sailors, and people living in the coastal areas have increased immensely. The size of ships and the number of offshore structures have grown into a web of international commerce, that influences all humans. These changes now affect our urban planning, large-scale funding, and political decision-making—all upon an undeniably rising sea level and changing climate. But still, by understanding

A surfer gets in position to take advantage of a peaking wave. Notice the shadows on the bottom, which reveal the texture of the water's surface. Kirra, Gold Coast, Australia. *Ted Grambeau* the origins of coastal dynamics, societies can anticipate coastal changes and respond to those changes in support of the well-being of their members. This edition of *Waves and Beaches* will help you take action and is in part why I became involved with its writing.

Satellites have deepened our understanding of the rapid changes affecting our coastlines. We now observe our planet's dynamic distribution of water, ice, and water vapor through distant eyes. Water has acted as a thermostat to keep our planet habitable, but our thermostat—comfortably stable for most of the past 4,000 years—has been reset. We know our climate history from the story told by a million years of atmospheric gases trapped in glacial ice in Greenland and Antarctica, the amounts of which reveal periods of warm and cold. Coral reefs, oceanic sediments, tree rings, and plant pollen data recount our ocean temperatures over the same chapter of history. These are irrefutable records of our planet's ancient sea levels, ocean currents, and atmospheric weather patterns.

These ancient dynamics have been disrupted by rapid population growth, the use of fossil fuels, the damming of rivers, the changing sediment loads, and the diversion of water flowing through estuaries and deltas as we contaminate aquifers with chemical pollutants. There is no alibi. The world's oceans are experiencing more intense hurricanes, rising sea levels, coastal erosion, storm surges, and saltwater intrusion into our freshwater aquifers. The waves and beaches of the world are elevating their battle during this swiftly transforming era, the most turbulent time since the "rise of humanity."

My awareness of the sea was early and multifaceted. By age five I had crossed the Pacific and Atlantic several times. I was bodysurfing by age nine, freediving at eleven, sailing at thirteen. Many of my teenage years were spent diving and exploring the same Pacific coast as Willard Bascom did in the 1940s. I have measured waves at sea from above, below, and within. I have spent many decades of my life aboard seagoing vessels, ashore examining beaches, all the while diving, surfing, and traveling the world's shores.

At universities in Europe (on the Baltic and Mediterranean Seas) I studied math and physics, then oceanography in graduate school where I encountered the first edition of *Waves and Beaches* as a textbook. My professional involvement with waves began with funding from the

Nearshore Sediment Transport Study (NSTS), which began where Bascom left off in the late 1970s. I became part of the "next wave" of coastal research in the 1980s. In the ensuing decades I plunged into the measurement of ocean turbulence, polar region science, oozing mud of the Amazon, instrument development, and global-scale changes.

In the latter 1990s Willard Bascom and I became friends. We spoke of diverse topics, sharing many interests including ancient history—bronzes from the age of Pericles—and waves. Indeed, as a father would interact with a son, he would give me "homework" to complete—he had not slowed down with age. Bascom would say, "Read this and tell me what you think." Or he would ask me, "How do they measure this now?" One day he handed me the second edition of *Waves and Beaches* and commanded: "Read this and tell me what it needs." That was the spark that marked the beginning of my transition from Bascom's student to being his collaborator. This edition, Bascom's last homework assignment for me, keeps the spirit of adventure alive while providing knowledge pertinent to the changes affecting us in the twenty-first century. It is my hope that this third edition of *Waves and Beaches* will help you take action and create your spark in the sea. — KM

straightens its front to present the least possible shoreline to the sea's onslaught.

When the great storm waves come, the beach will temporarily retreat, slyly deploying part of its material in a sandy underwater bar that forces the waves to break prematurely and spend their energies in futile foam and turbulence before they reach the main coast. When the storm subsides, the small waves that follow contritely return the sand to widen the beach again. Rarely can either of the antagonists claim a permanent victory.

This shifting battleground is the *surf zone* and the *beach face*. The two combatants continue their engagement in a coastal world that exists at the whim of a grander empire; now swept with the winds of climate change. The land will retreat before the rising sea levels with elusive transfers of sediments not seen for millennia. This is their story in the twenty-first century.

Units and Variables Used in This Edition

Units of measurement are fundamental to quantify anything. Here is an attempt to provide an understanding of the jumble of maritime-related units and variables found later in this book that may be difficult for some readers. The focus is to communicate concept, relative size, and magnitude. Today all scientific publications use the International System of Units (SI from the French *Système International*). In the maritime world, another unit system is inescapable: Imperial units, once used across the British Empire, is still used in the United States. Many international engineering efforts, such as the ill-fated Mars Orbiter, have stumbled upon remnants of these units. In this edition, Imperial units are used first, followed by SI units in parentheses. Whenever possible the conversion between units is approximated without diminishing significance.

As an example, the unit *ton* has multiple usages, but all are referred to in this book simply as "ton." There is a cornucopia of meanings for this unit. Ton can be used for weight (2,000 pounds, 1,000 kilograms, or 2,240 pounds for a "long ton") or volume such as freight (40 cubic feet), ship volume (100 cubic feet per registered ton), petroleum products (about 7 barrels of oil), and sand (0.6 cubic meters or 0.7 cubic yards). Understand the concept first, then read on, and if necessary, bypass a confusing jumble of units.

Units and conversions

ft	foot—Unit of distance, (0.3048 meters), there are 3 feet per yard (0.91 meters).
lb	pound—Unit of weight (0.45 kilograms, abbreviated kg).
kt	knot—Unit of speed (1.15 mph), (1.85 kilometers per hour, abbreviated km/h), (0.514 meters per second, abbreviated m/s), (514 centimeters per second, abbreviated cm/s).
cubic yard	Unit of volume 9 cubic feet, (0.76 cubic meters, also written m³).
t	ton—Unit of weight 2,000 pounds (metric ton is 1,000 kg) or volume (40 cubic feet) or (7 barrels of oil).
psi	pressure—Pounds per square inch, force per unit of area, 14.5 psi is equal to 1 bar.
J	Joule—Unit of work equal to a force times distance, also 1 Watt for 1 second.
W	Watt—Unit of power, equal to 1 Joule per second, also 1,000 Watts is equal to 1 kW.
kWh	kilowatt-hour—Unit of energy, power in kilowatts times the number of hours.
hp	horsepower—Unit of power (approximately 750 Watts).
F	degrees—Unit of temperature; to convert Fahrenheit to Celsius subtract 32 and multiple by 0.5555.

Variables

L	Wave length or distance between two points.
d	Water depth in units of distance (length).
H	Wave height measured from trough to crest (peak).
f	Frequency measured in cycles per unit time (same as $1/T$), one cycle per second is 1 Hertz (Hz).
g	Force of gravity, approximately 32.2 ft/sec^2 (9.8 m/sec^2).
π	(Greek letter "pi") approximated by 3.14 (ratio of a circle's circumference to its diameter).
T	A period of time, usually in seconds.
α	(Greek letter "alpha") a measure of angle in degrees.
C	Velocity expressed as a distance per unit time, (from the word "celerity" also written lowercase c).
cos	Cosine, trigonometric function.
tanh	Hyperbolic tangent, trigonometric function.
ρ	(Greek letter "rho") the density of a material in mass per unit volume.
K	A coefficient, typically used as a constant with an assigned value.
energy	In units of Joules, also in kilotons of TNT, which is equal to 4.184 terajoules.
power	Energy expended per unit time.

Everything is drifting, the whole ocean moves ceaselessly, just as shifting and transitory as human theories. —Fridtjof Nansen, Norwegian oceanographer, explorer, and statesman

1 Genesis of Land, Water, and Waves

Waves are undulating forms of energy that can propagate through any medium—wave energy travels throughout the universe. Waves may exist in solids or on the interface between any two fluids of different densities, but this book will deal primarily with those that travel on the surface between the ocean and the atmosphere. Although any kind of disturbance in the water is likely to generate waves, there are three prime natural causes: wind, earthquakes, and the gravitational pulls of the Moon and the Sun.

Wind waves are the most familiar kind; they are also the most variable and, in many ways, the most puzzling. The size and variety of the waves raised by the wind depend on three main factors: the velocity of the wind, the distance it blows across the water, and the length of time it blows. Moreover, the character of the waves changes markedly as they move away from the winds that created them.

The earthquake mechanism for wave creation is simpler and thankfully less frequent. In this version, a rapid motion of the subsea

PREVIOUS SPREAD: Water fractures and dissolves the solid Earth, as it has for billions of years. Pieman Heads, Takayna/Tarkine, Tasmania. *Andy Chisholm* rocks disturbs a mass of water. In regaining its equilibrium the water surface oscillates up and down and sends out a series of seismic sea waves, collectively called a *tsunami* (tsunami means "harbor wave" in Japanese from the characters *tsu*, meaning "harbor," and *nami*, meaning "wave").

The tides, which are a special kind of very long wave, are caused by the Earth's turning beneath great bulges of water raised by the combined rotational forces and gravitational fields of the Moon and the Sun.

The wave of climate change is another very long wave; it traverses human generations yet affects the power and location of winds, the genesis of waves, and the swash of every beach. We have just begun to see and understand its profound influence on all other waves. It is the most important wave ever experienced by civilization and one that cannot be separated from any conversation about ocean dynamics.

Regardless of the mechanism by which a wave is generated, the character of waves and the velocity at which they move are influenced by the depth of the water in which they are traveling. Therefore, in order to understand the behavior of waves, one must also know something about the shapes of the rocky basins that hold the water. So, let's first consider the beginnings of the water and the land—the origin of the Earth's surface features.

The Earth and Its Waters

The Earth formed from rocky and metallic fragments during the construction of the solar system—debris that was swept up by an initial nucleus and attracted together into a single body by the force of gravity. The original materials were cold as outer space and dry as dust; whatever water and gases they contained were locked inside individual fragments as chemical compounds. As the fragments joined, the Earth's gravity increased, attracting larger and larger objects to impact the Earth. This increasing gravity, combined with the timeless radioactive decay of elements like uranium and thorium, caused the new Earth to heat up. The internal temperature was such that many compounds broke down, releasing their water and gases. Plastic flow could occur. Segregation by density began, and the Earth started to organize into its

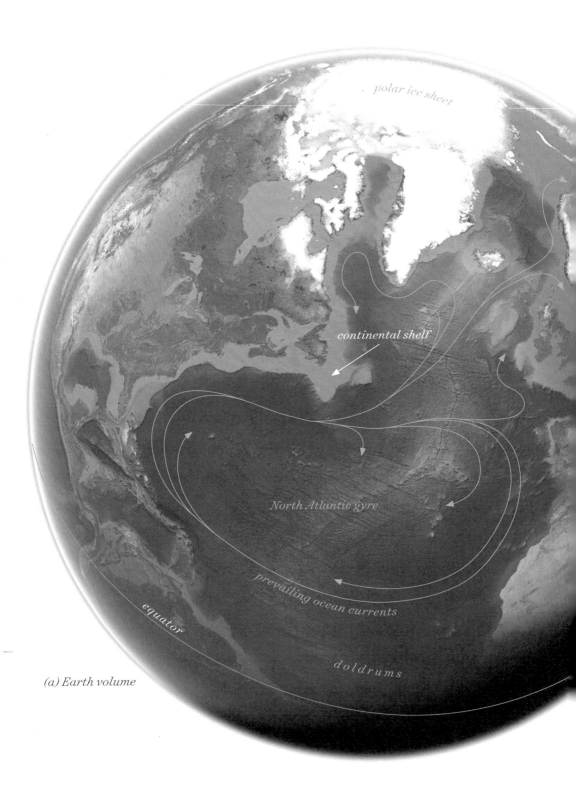

polar ice sheet

continental shelf

North Atlantic gyre

prevailing ocean currents

equator

doldrums

(a) Earth volume

present layered structure. The heaviest metals sank to the center; the lightest materials migrated outward.

The massive heat in the Earth led to motions of its rocky interior, much like the convection cell patterns seen in a boiling pot of soup. In turn, this process led to plate tectonics, a process where a conveyor belt of basaltic lowlands formed as the upwelling cell reached the surface, forming the ocean basins. As the basaltic plates dove down into the Earth's interior, they also caused melting, leading to the formation of the lightest rocks—granites—that reached the surface and collected over time into the large blocks now known as continents. At the same time, super-heated water and gases were brought to the surface by volcanic activity. The hot, steamy atmosphere cooled and condensed into liquid water, which flowed into low-lying basins. After a few billion years a global ocean had formed, and the atmosphere was sufficiently dense enough that effective winds could exist to transport the ocean's water vapor over the land. As soon as the evaporation-condensation cycle (hydrological cycle) could operate, rains fell, and stream erosion began. During one wave of cooling (a billion or so years ago), solid water—snow and ice—appeared as glaciers on mountains and ice caps at the poles. Fragments of continental rock were carried downhill by the running water and deposited into the ocean. In colder regions, glaciers ground away at the underlying rocks and provided fine sediments to the flowing waters below. The coarser particles were deposited close to shore; the finer ones were carried out to deep water, where they formed sedimentary

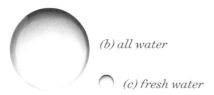

(b) all water

(c) fresh water

FIGURE 1: The Earth has a surprisingly small amount of liquid water. The liquid sphere (b) represents the amount of all water that circulates in the atmosphere, oceans, lakes, rivers, and groundwater of which 97 percent is in the ocean. The remaining tiny sphere (c) represents fresh water, only 3 percent of all water, for crops, drinking, and industry.

deposits that tended to smooth the seafloor and raise the sea level. The motions of the new atmosphere created the first wind waves, and these waves began the attack on the primordial shorelines. Just as they do now, the waves undermined sea cliffs, bringing down large chunks of rock, which were ground against each other by the moving water to form sand. The sand mined from the cliffs and the sand mined inland by streams were intermingled, sorted by the movement of the water, and redistributed along the shore. The first beaches formed.

As these processes proceeded over millions of years (the segregation of materials in the Earth's interior plate tectonic motions and new water arriving at the surface still occurs), the level of the ocean rose above the edge of its prior natural basin. The prior edges of the continental blocks have been flooded to an average depth of 600 feet (200 m), causing many shorelines today to be sandy and rocky (see figure 2).

It is well to remember that although the shoreline is important as the place where land and water meet, it is not the rim of the ocean in the geological sense. The true ocean basin begins well offshore where the edge of the continental rock slopes steeply into the watery abyss. In the basin the average water depth is nearly 15,000 feet (4,500 m) and the great waves race along at high speeds; on the shallow shelves these same waves are slowed by the drag of the bottom. Therefore, it is on the shallow continental shelves that many of the phenomena described in this book occur. On these shallow shelves, the waves moving landward from the deep ocean are transformed, where they first feel the bottom. It is here where beaches are created and constantly rearranged; where human constructions must meet and resist the force of the ocean's waves.

When viewed from space, the surface of Earth appears as primarily liquid, but do not be deceived. There is surprisingly little water on Earth in comparison to Earth's total volume. Of all the water on Earth, about 97 percent is in the ocean (see figure 1, page 22). The forces of nature engage the Earth's water as a weapon in a constant battle with the land.

The Sun's radiation warms the Earth, forming the atmospheric and oceanic circulation patterns. Once in motion, all are influenced by the Earth's rotation. Warmed tropical air rises and is replaced by cooler air from the north or south. This movement, driven by the heat of the Sun

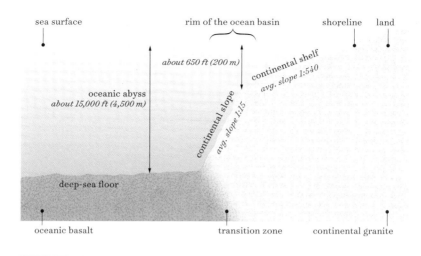

FIGURE 2: The contact between ocean, seafloor, and continent.

and guided by the rotation of the Earth, causes the major winds. Near the equator the air is relatively still (called the *doldrums*), but not far to its north or south the trade winds blow steadily to the west. At higher latitudes (40 to 50 degrees) the winds blow to the east.

These winds raise waves and provide most of the driving force for the great currents of the Earth. The trade winds give rise to the equatorial currents, flowing close to the surface toward the west until they encounter a land mass that turns them away from the equator. The water masses flow to the north or south and eventually close the loop, forming huge eddies, or *gyres*, that rotate clockwise in the Northern Hemisphere and counterclockwise in the Southern Hemisphere. Each ocean has these great rivers of water: those in the Atlantic being known as the Gulf Stream (North America) and Benguela Current (Africa); in the Pacific they are the Kuroshio Current (Japan) and the Humboldt Current (Peru-Chile). The smaller Indian Ocean basin has complex ocean currents that reverse their directions (between the Arabian Sea and the Bay of Bengal) during monsoonal conditions.

Waves come in many kinds and sizes, and for that reason it is best to think of them as a continuous spectrum extending from waves so small that they can hardly be seen to waves so long they are unnoticed in the period of a human lifetime. These subtle ebbs and flows of energy change our climate and affect the polar ice caps, sea level, weather patterns, and global winds; the essence of wave formation. Wave and beach processes only exist with the flow of energy. And today humans are influencing the Earth's energy flows and climate—its seasons, ice caps, storms, and winds, its sediments and sea level—we have become part of the spectrum.

THE WAVE SPECTRUM

Waves range in size from the short ripples in a pond to the great storm waves of the ocean and the tides, whose wave length is half the distance around the Earth. In order to be able to discuss such widely varying kinds and sizes of waves, it is necessary to agree on a standard set of names for the parts of a wave (see figure 3).

The principal parts are defined as follows:

Crest:	The high point of a wave.
Trough:	The low point of a wave.
Wave height:	Vertical distance from trough to crest.
Wave length:	Horizontal distance between adjacent crests.
Wave period:	The time in seconds for a wave crest to traverse a distance equal to one wave length.

There is a direct relationship between wave period and wave length, but wave height is independent of either.

Waves are classified according to their period; most range from less than one second to minutes (tsunamis) to hours (tides). Each undulation of each wave changes sea level for a characteristic period of time. Occurrences measured in years such as El Niño–Southern Oscillation (ENSO) bring storm waves, and the thousands of years–long period Milankovitch cycles (Earth's tilt and orbit patterns) affect sea level. The wave spectrum diagram (see figure 4, page 28) shows that the waves in the ocean are distributed among several major types, each with its characteristic range of periods and influence on sea level.

Beginning near the left side of the spectrum with the very short-period waves, we have in order: ripples, with periods of fractional

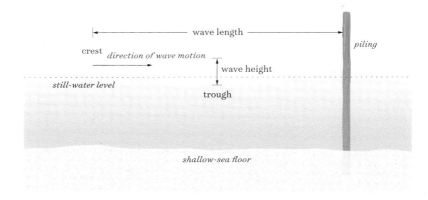

FIGURE 3: The parts of a wave. The period of the wave is the time in seconds for two successive crests to pass a fixed point, such as a piling. Wave height is from crest to trough.

seconds; wind chop, of one to four seconds; fully developed seas, five to twelve seconds; swell, six to twenty-two seconds; surf beat, of about one to three minutes; tsunamis, of ten to twenty minutes; and tides, with periods near twelve or twenty-four hours. Thus, there are many kinds of waves, each generated and developed in a special way.

Note that all the water waves just mentioned are called *gravity waves* because, once they are created, gravity is the force that drives them, by attempting to restore the original flat-water surface.

Each gravity wave is made up of two parts: the crest that rises above the average sea level and the trough that extends below it. As a group of waves moves over the surface of the water, each crest seems to be forever attempting to overtake the trough ahead, fill it in, and restore equilibrium. The wave source, whatever it was, worked against gravity.

One special form of wave not driven by gravity is possibly the most abundant kind of wave on the sea. The first tiny ripples that a light breeze raises on a glassy sea surface, or on the slopes of larger waves, are called *capillary waves*—capillary because they are controlled by *surface tension* and respond to the same forces that cause water to rise in capillary (very small diameter) glass tubing. The capillary force inside a small glass tube is stronger than gravity, so the water moves slightly upward.

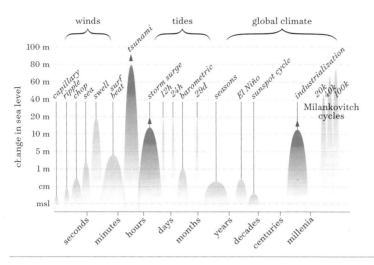

FIGURE 4: The ocean-wave spectrum extends in time from seconds to millennia. Each wave changes sea level during its own period. The most difficult waves to predict and quantify are shown in orange.

Surface tension is a property of liquids that makes them want to pull together and act as though they were covered by an elastic film. This force pulls water molecules together (cohesion) and determines the shape of each capillary wave and form of every drop of water. Capillary waves are often 1 or 2 mm high, a few centimeters from crest to crest, and are usually seen in groups of a dozen or so. In the sea, they arise when the drag of moving air stretches the surface and wrinkles the uppermost thin layer of water where there is a more systematic alignment of the molecules of water. Thus the size, slope, and velocity of these tiny waves are governed by the elasticity or tension of the surface film.

As might be expected in a phenomenon dominated by surface tension, the wave crests of capillary waves are rounded rather than peaked. A rise in the sea temperature will cause surface tension to reduce and viscosity to decrease, and these changes affect how waves are formed.

GRAVITY WAVES Long-period swell can travel great distances, far beyond where the waves were created. These gravity waves cross oceans with little loss of energy before reaching land. Raglan Township, New Zealand. *Rambo Estrada*

CAPILLARY WAVES Capillary waves are the smallest of waves, formed by the wind and objects, which stretch and wrinkle the water's surface only for it to be pulled back by surface tension. *Kim McCoy*

Unlike gravity waves, the shorter wave lengths move faster. Capillary waves give way to the development of ripples, also caused by minute pressure differences, which are at the beginning of our wave spectrum and lead to the growth of larger waves.

The simultaneous existence of so many kinds and sizes of waves on the surface of the ocean, coming from different sources, moving in many directions, and changing inexplicably from day to day, made it difficult for us to learn the ways of waves.

For example, see what happens when you toss several pebbles into a pool of water (pictured on page 31). The impulse generates a series of similar waves that move outward in all directions. The simple circular pattern is clear until the first waves reach shore and are then reflected backward. Now the pattern is not so simple, for the wave fronts of the returning waves interfere with the outgoing waves. The two sets of waves form curious patterns with diamond-shaped high

Waves radiate outward from their source until encountering other waves or objects. *YAY Media AS/Alamy Stock Photo*

points where crests coincide. As the reflections from the other sides of the puddle are added, the interference pattern becomes very complex. For a few moments there is a hopeless jumble of high points moving in all directions, and then the whole surface flattens back to mirror-like calm. You could perform this seemingly simple experiment a hundred times and still not clearly understand what happened.

In the ocean, however, the situation is far more complicated. First, the source of the waves is rarely an impulse at one point—usually it is a gusty wind blowing over a broad area that creates very irregular wave shapes. Second, waves change in character as they leave the generating area and travel long distances. Third, usually several sets of waves with different periods and directions are present at the same time. Fourth, waves are greatly influenced by the undersea topography (frequently called *bathymetry*). When they approach shore and move into shallow water, the wave fronts bend and the waves break, expending their energy in foam and turbulence. Plain and simple, any questions about the manner in which waves are born, develop, travel, or die cannot be answered easily by casual observers. In fact, ocean waves are so hopelessly complex that thousands of years of observations produced only the obvious explanation that, somehow, waves are raised by the wind. The stronger the wind, the bigger the waves, of course.

The description of the sea surface remained in the province of an anonymous poet, who found it "... troubled, unsettled, restless. Purring with ripples under the caress of a breeze, flying into scattered billows before the torment of a storm and flung as raging surf against the land; heaving with tides breathed by the sleeping giant beneath." A fanciful but quite useless description of the wave spectrum.

Now, after more than a hundred years of scientific work, including concentrated efforts since the 1940s, most of the major features of waves and their causes can be satisfactorily explained in mathematical terms and reproduced experimentally. Theoreticians use complicated equations and occasionally the study of waves slips into the hands of those who have never been to sea. Here we will resist such complexity and remain descriptive when possible.

THE EDGE OF THE LAND

The crust of the Earth is slowly but constantly shifting; the continents act much like great rafts of rock, floating on the viscous interior of the Earth and reshaped by plate tectonics. Consequently, if a load is added to the top of the raft—by a huge volcanic outpouring of lava or the accumulation of a great mass of ice, for example—the raft will sink a little and the sea level will appear to rise. By the same reasoning, as erosion removes land mass and large ice sheets melt, the load is lightened and the land rises. This rising of the land as ice sheets melt is referred to as post-glacial rebound. For example, a number of embayments on the Alaska coast that were used as harbors a century ago are now too shallow to be navigable because that part of the continent has risen. Seventy million years ago England's famous White Cliffs of Dover were a shallow seafloor. Many types of forces are at work—everything changes.

Other forces deep in the Earth also cause the great blocks of continental rock to move up and down and the ocean level to rise and fall. These major crustal movements occur very slowly, but as they do, the shoreline—which is especially sensitive to such changes—advances and retreats. Many geologists classify coasts according to whether they are submerging or emerging from the sea and whether erosion of rock or deposition of sediment has the upper hand. For example, much of the central California coast is rising, from Monterey to Mendocino. This movement is evidenced by the existence of terrace-like remnants of old sea bottom now well above sea level. Along the Northwest coast (i.e., Puget Sound) and Northeast coast (i.e., Hudson River Valley to Maine), large segments are described as *drowned topography*, meaning that the land has sunk relative to sea level. Because the original topography was largely hills and valleys, in both of these areas the shoreline is very irregular. Beaches tend to be narrow, short, and rocky; they do not form an important part of the coast.

Most of the East Coast from New Jersey to Florida is nearly straight because the submerged land has a long gentle slope that extends from many miles inland to the edge of the continental shelf 100 miles (160 km) offshore. Neglecting the rapid changes of the past century, this coast is stable, as is the coast of the Netherlands, and has not changed its elevation with respect to the ocean for a long period of time. Such coastlines are characterized by an almost continuous line of sandy barrier islands with great, wide beaches. Between these elongated islands and the mainland is a series of shallow bays and lagoons. So the basic shape, or the *geomorphology*, of a coast is the result of the ocean's history of interactions with the land. If there is an ample sand supply and if enough time elapses without a major change in elevation, the beach will become an influential part of the coast.

Most coasts have a rather complicated geologic history. Relative to sea level, they have at various times emerged and submerged again, each time retaining some features left from the previous iteration. Moreover, because the marine processes are usually interrupted before they are complete, there are relatively few examples of "finished" work. The geologist is thus forced to observe changing situations and guess how the process started and what forms it will eventually produce.

The principal concern becomes determining the mechanism that causes the changes and the rate at which the changes are taking place. Then, perhaps, the future can be forecast.

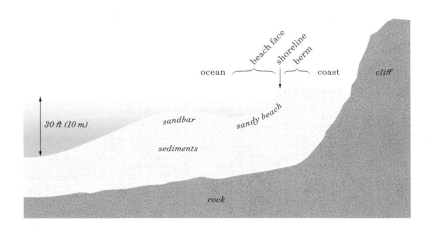

FIGURE 5: The anatomy of a beach, showing the locations of coast, berm, shoreline, beach face, and sandbar.

Before moving on, it's important to define three terms—shoreline, coast, and beach—that often cause confusion. A *shoreline* is the line of contact between water and land. A *coast* is a large physical geographic feature often extending several miles inland from the shore and several hundred miles along it. By comparison a *beach* is a relatively small feature of the coast, near the water's edge, whose limits are defined by the effects of waves (see figure 5).

A beach is an accumulation of rock fragments subject to movement by ordinary wave action. A tide will extend the process up and down a beach face. Beaches may be composed of any kind or color of rocky material, ranging in size from boulders to fine sand. Because most of the beach material along the most heavily populated part of the US coast consists of a light-colored sand—which is created as a result of the weathering of granitic rock into its two main constituents, quartz and feldspar—most of us tend to think of beaches as stretches of white

In Baja California each high tide erodes the cliff, permitting endless waves to churn cobbles into sand. Baja California, Mexico. *Kim McCoy*

sand. Some white beaches in Florida are made of finely sorted quartz sand as in Destin, Florida, whereas others are composed of carbonates: by-products from the waves grinding up coral reefs, shells, and other organisms. But many Pacific island beaches are made of black sand, formed by the disintegration of dark volcanic rocks. Many English beaches are composed of small flat stones called shingle, formed from the destruction of sea cliffs made of sedimentary rock, and many Alaska beaches consist of large cobbles. And for a hundred miles along the coast of Baja California, Mexico, the beach (see photo above) is made of two materials: a flat sandy portion that is exposed only at low tide, while immediately above and behind the sand, great cobble ramparts rise to a height of 30 feet (10 m) or more. Our idea of a beach depends on what we have been exposed to. In this book, for convenience, all beach material will be called *sand*, although it is recognized that all the features described may be formed in pebbles or shingle or cobbles.

We will not, however, explore in detail the complex study of sand, its composition, grain sizes, and mechanisms of formation, as it is outside of the scope of this book.

Moving ahead, we can think about beaches in two ways: (1) as small closed systems in which the sand moves either onshore or offshore at the whim of the waves, or alongshore in accordance with currents; and (2) as geologic units of considerable size.

Beaches as Major Coastal Features

Beaches are often of grand enough scale to be worthy of study as major coastal features. Although the comprehensive consideration of beaches in a physical geographic, or *physiographic*, sense is also beyond the scope of this book, it is important to briefly consider the three forms that beaches are most likely to take when they are treated as geologic units. A beach can be simply a narrow strip of sand separating the rocky cliffs of land from the sea; a spit or a baymouth bar; or a barrier island.

The first form—a beach that is narrow, of limited extent, and on which the sand is a shallow veneer over the rock—is indicative of a youthful shoreline. That is, not much time has passed, geologically speaking, since the most recent change in sea level. What little sand there is has been created in place by the undermining of the cliff and the grinding of the rocks by wave action. Many beaches of the California and Oregon coasts are in this category. These so-called *pocket beaches* (see photo on page 37) extend between rocky headlands and often have sheer cliffs behind; in the winter months, storms strip off most of the sand, exposing cobbles and the underlying rocks.

The second form, in which spits and baymouth bars are created by wave action, requires more time to develop. In it, rough coasts tend to be straightened by wave forces and ragged shorelines are smoothed. Headlands extending into the sea are attacked because wave energy is focused on them by the underwater topography. Waves striking the coast at an angle create longshore currents that transport sand, and seal off the mouths of relatively quiet bays.

The sequence of events in one form of coastal straightening is illustrated in figure 6 (page 38). At stage one, bold headlands project into the

POCKET BEACHES Pocket beaches form between rocky
headlands. Nusa Penida, Bali. *Tommy Schultz*

ocean where they are attacked by waves whose energy is concentrated
there by the process of wave refraction. As the headland retreats and
the cliffs are reduced to rocky fragments, currents caused by the waves
striking the shore obliquely transport the smaller particles into the
relatively quiet water at the head of the bay where they form a beach,
stage two. Later, the headlands have been cut back and the bay becomes shallow, as in stage three. The longshore coastal currents, which
were disorganized by turbulence around the headlands in the earlier
stages, now become dominant and sweep sand along the coast, creating
beaches and baymouth bars (see photo on page 39). With a straight
shoreline, sand can be transported considerable distances, passing
headlands and bays alike in its longshore migration. Eventually, at the
land's end, the water deepens and the transporting current spreads out
and is reduced in velocity so that the sand it has been carrying drops to
the bottom. These embankment-like deposits in which the outermost

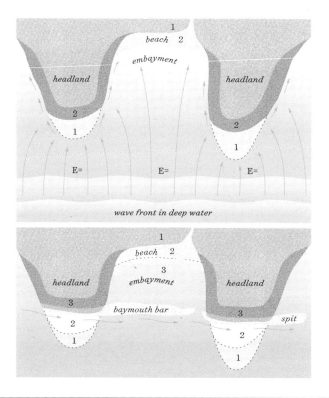

FIGURE 6: Waves straighten a rocky coast in stages. Top: Zones of equal wave energy in deep water are concentrated by wave refraction so that the headlands are attacked. Bottom: Eventually headlands are cut back to furnish enough sand to build beaches, bars, and spits.

end is surrounded by water are called spits. They form wherever there is a supply of sand, a transporting current, and a dumping ground.

There are two famous spits at the entrance to New York Harbor (see images on pages 40 and 41). Sandy Hook, to the south, was built by materials supplied by the erosion and retreat of the Navesink Highlands in New Jersey. It grew steadily until it reached an equilibrium situation in which the new sand added to the tip is just equal to that removed by the tidal currents at the harbor mouth.

Rockaway spit, northeast of the harbor entrance, was built with sand from the Long Island coast and grew at the rate of 200 feet (60 m) per year (1 mile in twenty-three years) for a long period until the present series of groins and jetties were built. Frequently these rivers of sand

BAYMOUTH BAR AND SPIT Baymouth bars are formed by
the deposition of sand, sometimes enclosing an embayment. A spit
is similar, but is attached to land at only one end. Klamath River,
California. *Gary Crabbe*

flowing along a coast are supplied by the erosion of valuable property.
This erosion creates one form of a beach problem; later the sand is deposited where it is not wanted, still another problem.

The beaches of the north Pacific coast are composed of an abundance of fine dark sand made from the disintegration of an inland
basaltic plateau and brought to the sea by the Columbia River. The
sand is distributed by wave and current action so that both north
and south of the river mouth great spits have formed, straightening
the coast by sealing off bays and headlands. These spits are continually widening, as evidenced by a series of sandy ridges or growth
lines, and the underwater sandbars opposite bay entrances are constantly shifting.

NEW YORK BAY, 1860 Two large spits (Rockaway and Sandy Hook) at the New York Harbor entrance have changed due to the ongoing work of erosion and deposition since this chart was made in 1860. New York. *US Coast Survey, Lindenkohl & P. Witzel*

There are large meandering sandspits in Europe, at the mouth of the Bay of Arcachon, near Bordeaux, France. There the winds blow sandspits into Europe's largest dunes—over 330 feet (100 m) high. The observer, feeling the shudder of the beach and hearing the roaring of the great winter breakers, gets an impression of natural forces in violent conflict, and one wonders why the changes are not more rapid.

The third major beach form, the *barrier island*, makes up a major part of the East and Gulf Coasts of the United States and much of the coasts of Holland and Poland. A half dozen major cities are built on these sandy strips, including Atlantic City, New Jersey; Miami Beach, Florida; and Galveston, Texas. Sometimes called barrier beaches or even offshore bars, these islands vary in width from a few yards to a mile (1.6 km). They may be dozens of miles long and are, in

NEW YORK BAY, 2020 Rockaway Spit was ravaged by
Hurricane Sandy in 2012. Today sand-nourishment and dune-
restoration projects hope to resist the effects of climate change.
New York Harbor, New York. *NASA/Landsat*

places, separated from the mainland by shallow bays many miles wide
(see photo on page 42).

When sand is blown by the wind into dunes, as at Kitty Hawk, North
Carolina (where the Wright brothers first flew in 1903), the hills on
the islands may rise to a height of nearly 100 feet (30 m). Dunes act as
a reservoir, a buffer of sand held in reserve for extreme events. When
human activities compromise the dunes' ability to serve this function,
the natural supply of sand is decreased and the beach is starved of sand
and coastal erosion advances. We must respect the dunes.

Frequently between the dunes and the main coast there is a chain
of bays, marshes, and tidal lagoons, which in many places have been
developed into an inland waterway where small craft can move safely
along the coast. These large, sandy shoreline features are accumulated

BARRIER ISLANDS Barrier islands protect the mainland from
storms and shelter inland bays, shallow marshes, and tidal lagoons.
Cape Lookout, North Carolina. *Steve Dunwell*

beach deposits which have grown so large and permanent that they no
longer fit our definition that limits a beach to the area in which the sand
is moved by ordinary wave action. In some areas such as in Namibia, on
the west coast of Africa, coastal dunes rise to over 1,000 feet (300 m)
and then extend their complexity almost endlessly into the desert.

Since beaches owe their existence to wave action, they have a dynamic
yet chaotic quality. That is, beach materials are always in motion—as
long as there are waves—although this complex mobility is not readily
apparent to the casual observer. The motion of the beach material may be
parallel to the shoreline, in which case it is transported by alongshore currents, or it may be moved toward or away from the land by wave action.

There are two major beach forms created by the waves: *berms* and
bars. Berms are flat, above-water features that make up the familiar part
of the beach. Bars are underwater ridges of sand that parallel the shoreline and are seldom seen except at unusually low tides. On most beaches

BEACH SAND DUNES Dunes with vegetation hold sand in reserve for extreme events. Along the shore a sandbar is exposed at low tide. Dunes and bars work together to buffer the coastline. Jekyll Island, Georgia. *Kim McCoy*

there is a constant exchange of sand between these two features, and the direction of the transport depends on the character of the waves. When the waves are large and follow each other closely (as they do under storm conditions, for example) the berm is eroded and as a result, the offshore bar builds up. When calm conditions return, the small waves rebuild the berm at the expense of the bar. For this reason, the above-water part of a beach is generally much narrower in the stormy winter months than in the summer, which is convenient for the hordes of bathers who come to sun themselves on the wide summer berm and swim in the low surf.

The steeply sloping seaward side of the berm against which the waves are in constant contact is called the *beach face*. The face might also be described as the zone within which the shoreline wanders as the waves rush up the beach and wash back down it. This is the *swash zone*.

It is necessary to set limits on the extent of what is considered a beach, to keep the discussion of its properties within reasonable bounds.

The coastal sand dunes of Namibia contain diamond-rich sediments from the Orange River that have been transported northward by the wave-driven longshore currents. Namib-Naukluft National Park, Namibia. *Fabian von Poser/Getty Images*

In the seaward direction, beaches extend outward as far as where ordinary waves move the sand particles. This limit has been found to be about 30 feet (10 m) below the low-tide level (see figure 5, page 34). This is an arbitrary but satisfactory limit that is generally accepted. Above water, the beach extends landward to the edge of the permanent coast. The latter may consist of a cliff, sand dunes, or human-made structures. In the geological sense these are not really permanent, but they endure far longer than the small-scale beach features that concern us here. The Earth's coastlines have not always been the way they are now. Geologically speaking, most shorelines and beaches are very recent features. The present sea level has existed with little change, changing only a few feet over the past 4,000 years. This is because we have been living in an interglacial calm after a period when great ice sheets covered the Earth and sea level was much lower. Twenty thousand years ago the global sea level was also fairly stable—but about

SAND BERMS Berms are generally flat, above-water features formed by waves during high tides. Berms eroded by steeper storm waves can reform into underwater bars. Salt House Beach, Norfolk, United Kingdom. *Anthony Bennett*

400 feet (120 m) lower than today. The Mediterranean island nation of Malta was connected to Italy; now it is 60 miles offshore. However, by roughly 18,000 years ago, a warming period had been set in motion by the Milankovitch cycles. These are changing patterns in the Earth's tilt and orbital movement around the Sun (roughly 20,000, 40,000, and 100,000 years) that influence the distribution of the Sun's energy reaching the Earth. So, as ice sheets began to melt more rapidly, the ocean temperatures increased and sea levels rose (seawater expands with increasing temperature), changing the location of Earth's water and the dynamics of Earth's coastlines. During the Younger Dryas cold snap about 13,000 years ago, when pulses of cold meltwater went into the North Atlantic, the rate of sea-level rise slowed for a millennium, lessening the sea's rate of assault on beaches. But the Earth continued to warm, glaciers continued to melt, and sea level rose for many thousands of years, creating new shorelines and beaches. The "old" beaches

SANDBAR Bars are underwater ridges of sand that parallel the shoreline and are seen only at low tide. Water, elevated by waves, flows over the bar and then parallel to the shore, forming sand ripples on its way back to the sea. Shackleford Banks, North Carolina. *Kim McCoy*

were left under hundreds of feet of water. No beach on Earth was left unchanged. Although sea level has been comfortably stable for most of the past four thousand years, previous sea-level rises provide earnest proxies for how today's changing climate is already affecting the dynamics of coastal regions.

In some areas, geological (or, tectonic) uplift and the post-glacial rebound have caused the land to rise relative to sea level. This means that few shorelines have reached steady-state conditions. An example of this post-glacial rebound is the city of Pisa, Italy, which was a port less than a thousand years ago. Even though the city was never covered by glaciers, it is now over 2 miles (3 km) from the sea. Excavations in the ancient city of Ephesus on the coast of Turkey have uncovered a colonnaded marble road from the harbor where distinguished visitors in Roman times could make a triumphal entry after landing. Now, the point of debarkation is more than a mile (2 km) from the sea. Such localized and regional post-glacial rebound changes will, of course, continue. Today, scientific measurements indicate that a new wave of global sea-level change began with the onset of the Industrial Revolution (see figure 4, page 28). It has sparked a new battle between the land and the sea.

The Longest Wave

We are experiencing a wave of climate change that began within the past two centuries. Our liquid oceans and gaseous atmosphere now flow with a new wave of energy enveloping our planet. Water stabilizes our climate. It takes energy to change rigid ice into flowing water. It takes even more energy to change liquid water into water vapor. We know that the flow of energy connects everything. It influences the formation of continents, earthquakes, volcanic eruptions, the strength of the Sun's rays, the amount of ice, sea level, human population dynamics, types of atmospheric gases, ocean circulation patterns, seasons, the warm-cold meanders of the jet stream, winds, ocean waves, beach erosion, and sand formation. All are intertwined and essential.

We have disrupted the flow of energy to power our civilization. Per capita energy usage has increased one-hundred-fold in the past 1,000 years and doubled in the twentieth century. There is a direct relationship between the gross domestic product (GDP) per person of a country and its per person energy use. A wave of global population has upset the Earth's climate and the resulting wave of climate change is now upon a densely populated coastline.

2   Ideal Waves

The shape and motion of the ocean's surface as waves pass across it is very complicated. It is little wonder that even after thousands of years of observation, seafarers developed no satisfactory explanation of the mechanics of wave motion. Ancient mariners knew in a general way that waves were generated by the winds, that they continued to travel outside the storm area, and that upon entering shallow water they would rise up and break, expending their energy on a beach or against a rocky headland. These characteristics were easily observable. Since ancient times the Polynesians have navigated vast distances aided by their knowledge of how waves interact with islands, are affected by currents, and propagate in the open ocean. They do not use any mathematics or computers, but they understand how waves can be refracted, reflected, and shoal, subjects that will be covered later in this book.

A major difficulty in explaining their origin and motion came from the fact that waves are so irregular. In the past, shipboard observers could see that when a breeze would suddenly spring up, a previously

PREVIOUS SPREAD: Distant winds created these waves. Each wave crest moves the water in the direction the wave is traveling, then each trough moves it back, close to where it originally was. Many types of waves exist, even in the clouds above. Milford Sound, New Zealand. *Matt Dunbar* calm sea would first become rippled, and then in time and as the wind increased, these ripples would grow into larger and larger waves. Soon the ship would be surrounded by a full-fledged storm with large irregular masses of water moving on all sides, often breaking on the deck. There was no longer any chance to observe—the goal was to survive. Away from the wave-generating area the waves tended to be seen as somewhat more regular. These observations were confused by the simultaneous existence of several kinds of waves from different storms and by the curious effect of the underwater topography.

Observers on shore would see high waves intermixed with low ones; several would arrive in quick succession, and then the time between waves would be long—some would break, others not. In the midst of a period of calm weather and blue skies, suddenly great waves would arrive at a shore, and no one on the shore could explain why. No general set of rules for wave behavior could be worked out that seemed to cover all the conditions observed. Even careful observers could only say, "That is just the nature of waves."

The obvious way to unpack such a complicated problem is to deal with each of the components, one at a time, in its simplest form. The first component of the problem is to define wave properties or dimensions and to determine the relationship between them. Next it would be necessary to discover what size and duration of storm and what velocity of wind created what kinds and sizes of waves. Finally, the relationship between wave motion and the depth of water needs to be worked out.

It seems easy now, with the advantage of hindsight and modern technology, to organize the wave-research program that could have been carried out many years ago, but, of course, nothing so systematic happened. Casual observations of many wave characteristics gradually led, step by step, to sufficient understanding that a beginning could be made on wave theory.

One can imagine that several of the important properties of waves were first thoughtfully noted many thousands of years ago by someone living on the shore of the sea. Probably this early scientist was regarded as the village slacker; probably making observations in a little cove with clear shallow water, a sandy bottom, and an occasional stalk

Not all waves are created equal. Irregular storm waves in the Antarctic engulf a vessel on all sides. The deck and chest-high rail are awash while 50-knot winds launch distant wave crests into the air. Drake Passage, Southern Ocean. *Kim McCoy*

of seaweed. The shore provided the vantage point to watch the waves of the sea move into the cove.

One bright and sunny day when no breeze blew to ruffle the surface of the water, a series of regular waves entered the cove, moved across it, and broke on the beach at its head. Idly this scientist-by-accident tossed a stick into the water and watched it float there, noting its position in relation to a rock on the opposite shore. It would rise and fall and move back and forth as waves passed under, but it did not move shoreward with the waves. There was nothing new or startling here; others had also seen this countless times before. But suddenly the idea occurred in the person's mind that waves are only moving forms and that the water stays in the same place. The stick, and the water around it, moved in a slow circular oscillation as each wave passed. The seaweed stalks were carefully observed growing upright from the bottom and moving in slow vertical circles. As a crest approached, the upper part of the weed moved toward it; as the crest passed, and the weed continued to point toward it until a new crest approached. Fragments of weed suspended in the water that were neither floating nor sinking moved in slow vertical circles, one for each passing wave. Here was a basic principle of

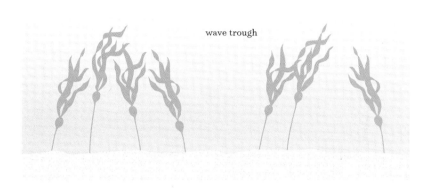

wave crest

wave trough

FIGURE 7: The motions of seaweed indicate the movement of water particles as waves pass.

hydrodynamics: Objects in the water tend to do what the water they displace would have done (see figure 7). The water particles also must be moving in circles.

With these simple early observations *oscillatory waves* were discovered. This was an accomplishment roughly equivalent to Newton's observation that an apple falls to the ground because of the force of gravity. Everyone had seen what it meant, but no one had thought about why. However, unlike Newton, early wave observers were not capable of expressing what they had seen in mathematical terms.

THE FIRST WAVE THEORY

In 1802, Franz von Gerstner (a German-Bohemian working in Prague) produced a nonlinear wave theory. He described how water particles in a wave move in circles, and he pointed out that those in the crest of

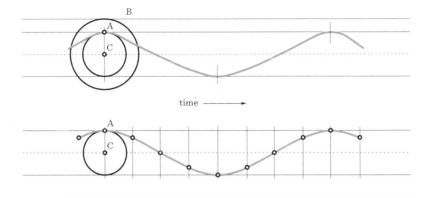

FIGURE 8: Geometrical waveforms for waves of equal lengths and heights but different shapes. Top: A trochoidal wave is generated by point A as the outer circle rolls along the underside of line B. Bottom: A sine wave is generated by projecting the position of point A to equal increments of time as it rotates about a stationary center C.

a wave move in the direction of wave advance, and those in the trough move in the opposite direction. Gerstner noted that before returning to its original position, each water particle at the surface traces a circular orbit, the diameter of which is exactly equal to the height of the passing wave. He observed that the surface trace of a wave is approximately a *trochoid*, the curve described by a point on a circle as the circle is rolled along the underside of a line (see figure 8). Presumably, he knew that if the wave height is small compared to the length, as it is for most water waves, the shape of the trochoid approaches that of a sine curve.

Such was the theoretical beginning. Gerstner's work was found later to have several inconsistencies, but it attracted the attention of the Weber brothers of Germany, Ernst and Wilhelm, who became the first wave science experimentalists. In 1825 they published their findings on using a glass-walled tank, which began the formal study of waves under controlled conditions. The flat-sided tank—which offered the opportunity to study one wave at a time at eye level and the chance to repeat an experiment over and over until understanding was achieved—overcame the major difficulties of studying waves in nature.

The Weber brothers discovered that waves are reflected from a vertical wall without loss of energy, and they watched suspended particles to confirm the theory that the circular orbits diminished in size with

WAVE CHANNEL A wave channel (wave flume) can create waves with controlled periods and heights to study wave forces on sediments, beach faces, buoys, and structures. *OSU O.H. Hinsdale Wave Reserch Laboratory*

increased depth. They found that, near the bottom of the tank, the orbits were greatly flattened.

These early theoreticians devised equations in which an endless train of perfect waves, all exactly alike, moved across an ocean of infinite breadth and depth. Their trains were an unreal abstraction, but the method was the most reasonable way to begin to work out the relationships between period and wave length and velocity for sinusoidal ocean waves. These equations were then applied to the waves observed in wave channels.

In the mid-1800s, there was another great flurry of experimental work. A Scotsman named John Scott Russell, for example, built a channel about 20 feet (6 m) long. Near one end was a removable gate that created a small reservoir when lowered in place. To make waves, Russell would suddenly raise the gate and allow the water to rush down the channel as a solitary wave or *wave of translation*. This impulse produced normal waves, which would then reflect back and forth between the ends. With this experiment, Russell had made the first careful measurements of wave velocity.

The development of equations and the serious attempt to understand theoretically what the waves in the tank and the ocean were doing was a big step. But we must remember that with equations the experimenter has true models in the wave channel; without equations the wave tank is only a plaything.

The theoretical and experimental work done today is more complicated because it is now commonly known that ocean waves are not really sinusoidal, or any other "pure" mathematical shape. Now, real ocean waves are dealt with *statistically*, as combinations of great numbers of small waves. However, model work today is done for the same reason that it was done long ago—to simplify the problems by working under controlled conditions. The Mediterranean cove of the past has been replaced by the experimental wave channel and the irregular swell by precision generators.

Today, many fluid mechanics laboratories have facilities for modeling waves and determining their effects on beaches and ships. These facilities range from the tabletop ripple-makers to huge tanks that can create breakers 8 feet high on full-size models. An example of a wave channel is shown in the photo on page 55.

The Fundamental Properties of Waves

Wave tanks allow us to change the wave period or wave height independently. One of the first things to be noticed is that the wave length depends upon wave period, whereas wave height does not. A wave's height is related to the amount of energy that created the wave. Several equations are introduced in the next paragraphs—hang on, like waves these too will soon pass.

Under ideal laboratory conditions, a pure sine wave can be produced. We discover that the wave length L (in feet) is equal to 5.12 times the square of the period.

$$L = \frac{g}{2\pi} T^2$$

Which can be simplified to $5.12\,T^2$ (or in meters $L = 1.56\,T^2$) and where L is the wave length in feet, g is the acceleration of gravity (32.2 feet per second, per second), and T is the period in seconds. Thus a one-second wave would be 5.12 feet (1.56 m) long; and a ten-second wave would be 512 feet (156 m) long. Wave velocity, the distance a wave travels per unit time, is usually designated by C (also known as "celerity"), and expressed as $C = \sqrt{\dfrac{gL}{2\pi}}$. A one-second wave verifies the relationship by moving at 5.1 feet per second (1.5 m/sec).

But, as the period is increased, one soon finds that these relationships for length and velocity do not hold exactly in this wave tank for periods of over one second. Why? Because in the shorter-period range we have been dealing with *deep-water waves*. A *shallow-water wave* is one that is traveling in water whose depth is less than half its wave length; that is, if the depth of water is small compared to the wave length, the effect of the bottom is sufficient to alter substantially the character of the waves. With the still-water depth in our tank at 2.5 feet (76 cm), increasing the period has produced waves which "feel the bottom" and are affected by it. The mathematical term "hyperbolic tangent" (tanh) might be difficult for some readers; however, the full expression for wave velocity, water depth (d) is taken into account:

$$C = \sqrt{\frac{gL}{2\pi}\,\tanh\frac{2\pi d}{L}}$$

The final term contains the ratio $\dfrac{d}{L}$ or *water depth/wave length*.

Most wave researchers find it convenient to describe waves in terms of their $\dfrac{d}{L}$ and use a simplified version of that equation. For a $\dfrac{d}{L}$ of more than 0.5 (a deep-water wave), the hyperbolic tangent of it is so close to 1 that it can be neglected, as we did earlier. On the other hand, if the depth is quite small compared to the wave length ($\dfrac{d}{L} = 0.05$) of the hyperbolic tangent; tanh $\dfrac{2\pi d}{L}$ can be replaced by simply $\dfrac{2\pi d}{L}$. Then, after cancellation, the wave-velocity expression becomes much more simple: $C = \sqrt{gd}$. This relationship is used for shallow-water waves.

This, too, is very convenient, especially when one is working with seismic sea waves, which are so long that, for them, even the deepest ocean is shallow water. But for values of $\dfrac{d}{L}$ between 0.05 and 0.5 we must use the longer form of the equation for deep-water waves.

We first thought that the wave height seemed to be independent of both the period and wave length, but further experiments show this is not quite so. If we hold the period constant at one second (with the wave length remaining 5.12 feet or 1.56 m) and gradually increase the wave height, we discover that waves higher than 0.75 feet (20 cm) have unstable crests. That is, they tend to break as they travel down the tank. On repeating the experiment with other heights and lengths we discover that the angle at a wave crest may not be smaller than 120 degrees, or the wave will break. Stated in another way, the wave height may not be greater than one-seventh of the wave length.

This relationship (ratio) of height to length ($\frac{H}{L}$) is called *wave steepness* (see figure 9).

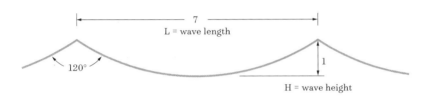

FIGURE 9: Maximum wave steepness is 1/7. Waves become unstable as the wave-crest angle decreases to less than 120 degrees.

In reviewing the notes on these past experiments, we find that as the waves became steeper they also increased slightly in speed until at 1:7, the maximum, they moved perhaps 10 percent faster than the theoretical speed. However, because ocean waves rarely achieve such steepness— only in violent storms—it is customary to neglect this increase.

When the waves move onto the abruptly shoaling beach at the end of the tank, they change character in another way. As the depth decreases, the waves are said to *peak up*; that is, their height increases rapidly. At the same time, the shallow water causes the wave length to decrease, and the result is a suddenly steepened wave. In a very short distance, the crest angle decreases below the critical 120 degrees and the wave becomes unstable. The crest, moving more rapidly than the water below, falls forward, and the waveform collapses into turbulent confusion, which uses up most of the wave's energy. Small amounts of this energy is converted into heat by frictional forces, just like rubbing your hands together. Trapped air produces sounds heard as the rumbling surf above the grinding noises of cobbles and the swishing of moving sand.

Stellar Nightwatch

Navigation has been revolutionized by the Global Positioning System (GPS), yet adequate older methods remain embedded in nature. In December of 2007, we sailed across the Atlantic guided by waves and stars. Our course was set southwest from the island of La Palma off the coast of Africa, across the Atlantic to the Caribbean some 2,500 nautical miles away. When beyond sight of land after night had fallen, the stars, planets, and waves guided us. The rising, culmination, and setting locations of stars and their visibility times provided us with heading and latitude. The long-period swell gave us direction and speed. Each surging wave dipped our bow, lifted our stern, and altered the vessel's course a few degrees back and forth while slowly passing under us. Above, the nighttime sky was our celestial compass. We kept the upright mast swaying between Mars and Jupiter, the luff of the mainsail followed the constellation Pleiades as the forestay pointed at a star low on the horizon. We forgot about the boat's compass; we didn't need it. Far from shipping lanes, our course was in the waves—each a cipher to our destination. All were there for us through the darkness until dawn. – KM

ORBITAL MOTION

Until now we have considered only the movement of the waveform along the surface of the water. Let us look inside the wave, at the motion of the individual water particles. What do they do as a wave passes? Since the water particles themselves cannot be seen, we must add a number of small markers to the water which will follow the water motions. The markers should be the same density as the water (neutrally buoyant), and easily visible.

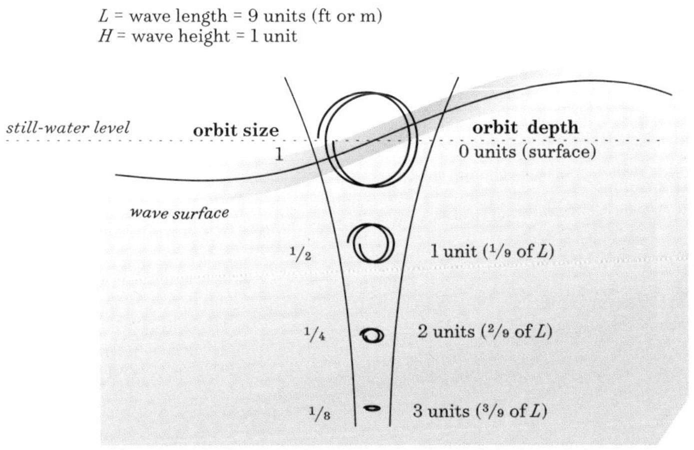

L = wave length = 9 units (ft or m)
H = wave height = 1 unit

still-water level   **orbit size**   **orbit depth**
                            1   0 units (surface)

wave surface

                         $^1/_2$ 1 unit ($^1/_9$ of L)

                         $^1/_4$ 2 units ($^2/_9$ of L)

                         $^1/_8$ 3 units ($^3/_9$ of L)

FIGURE 10: Wave orbital motions decrease in size and flatten the deeper they are below the surface. In water deeper than half of the wave length, wave motions are so small they are negligible, little energy is lost, and the wave is considered a deep-water wave.

The circular lines in figure 10 outline the orbits of four particles at four different depths. All the orbits of particles are in the same direction at the same time—they are in phase.

The surface particle has a circular orbit exactly equal to the height of the wave. The next one down made a somewhat smaller circle. The third orbit is not only smaller but is slightly flattened, and the one at the bottom moves back and forth in a straight line. By combining a series of such measurements with theoretical work, it has been established that at a depth of one-ninth the wave length, the diameter of the orbit is approximately halved.

For simplicity, let us use a wave with a wave length of 9 units (feet or meters) which has a wave height of 1 unit. At the surface, our wave has an orbit diameter of one unit. Below the surface at a depth of one-ninth the wave length (1 unit depth), it is found to have a particle orbit of one-half unit. Yet deeper, at a depth of 2 units, the movement is reduced to one-fourth the surface orbit diameter. At 3 units depth, the movement is mostly horizontal and diminished to one-eighth the size of the surface value. (All depths are relative to the still-water level, the water level in the absence of waves.)

After a series of a dozen or so waves have passed, the near-surface particles are seen to be describing circles that are not precisely in line with the first ones. So we trace out the new path and find that the new orbit is a little farther from the wave generator.

Mass Transport

This difference in wave position is the result of what is called *mass transport*. Until now we had supposed that the water returned to its original position after each wave passed, but then we found that when waves are steep, the orbital circles of the water particles do not exactly close. The water itself is transported by the passing waveform, although its progress is very slow compared to the wave velocity. The volume of water so moved is negligible for waves of small steepness (the usual circumstances) and can be disregarded for all practical purposes.

However, the existence of this mass transport is a serious matter to theoreticians. The early workers, including Gerstner and W. J. M. Rankine, concerned themselves only with *rotational flow*, in which each particle completed a perfect circle as the waveform passed. Later, George Airy and C. C. Stokes developed the *irrotational theory*, which requires that the water move forward slightly as the waveform passes. This irrotational flow of water due to wave action is now called *Stokes drift* and is in the direction of wave propagation.

This actual motion of the water is proportional to the square of the height of the waves, and is much more pronounced at the surface than a short distance down. In a wave channel, there is an imperceptibly small return flow of water along the bottom in the opposite direction to compensate for the surface transport by the waves.

One final experiment remains to be performed. We must now examine what happens to the water particles as the wave is transformed into a breaker by the sloping beach at the end of the tank (see figure 11, page 62). The breaking motion is too rapid to be followed. Moreover, each breaker tends to disintegrate and cast the particles up on the beach. A more sophisticated method of recording the wave motion is now possible by using multiple fast frame-rate synchronized cameras, feature-recognition software (similar to facial recognition), and proper lighting.

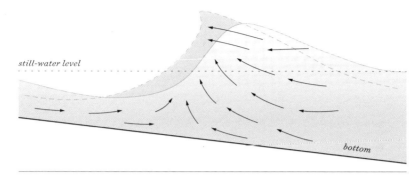

FIGURE 11: Movement of water particles as a wave transforms into a breaker in shallow water. Bottom friction causes the upper part of the wave to shear forward with all of its remaining energy.

The Forces of Gravity, Buoyancy, and Viscosity

It is important to keep in mind three fundamental forces that exist in every ocean wave, and form every beach. Gravity, buoyancy, and viscosity are three forces that struggle for dominance as they eternally flirt and dance with each other. Their presence is essential for the motion of waves and the paths of sediments.

Gravity is attractive; think of it as the universe's charisma. Gravity attempts to pull all matter toward a center, causing galaxies to form and comets to crash, propelling a surfer down the slope of a wave. Gravity is what causes tides to rise, waves to undulate, and a grain of sand to resist fluid motion. Amazingly, humans can describe gravity but have yet to agree on what causes it.

Buoyancy is an uplifting force. Think of buoyancy as lighthearted and youthful; it makes things lighter when the mood gets heavy. Buoyancy pushes back against gravity's pull. The buoyant force is what causes objects to float. It is asserted in all fluids, liquids, and gases alike. An object that is lighter than its surrounding fluid will move upward, whereas the object that is heavier will sink. When the buoyant force is equal to the gravitational force, an object becomes neutrally buoyant, drifting freely, neither up nor down. But it rarely lasts. Each grain of sediment in an offshore sandbar, every cobble, and every boulder gets buoyed by the uplifting force of its surrounding water.

Viscosity is the stickiness of the universe as energy spreads out. The viscous force comes forth as any fluid tries to resist flow. When there is movement, there is friction. The amount of friction in a fluid (i.e., viscous force) is linked to the *shear*; that is, when there are relative speeds within a fluid. Viscosity causes individual grains of sand to start moving beneath a wave, it mobilizes large stones, and is inherent in the drag forces that restrain the movements of tides, ships, and surfers. Viscous forces can be great enough to overwhelm both gravity and buoyancy.

Understanding the twirling dance of gravity, buoyancy, and viscosity is important. Together they influence a grand cascade of energy: the movement of planets, the heat from the Sun, the direction of atmospheric winds, the motion of ocean currents, and the formation of waves. These forces, all active within waves, even appear as the small-scale turbulent *saltation* (from *saltus*, "leap" in Latin) for each grain of sand. Our waves and beaches are the exhibition of a dynamic act, performed on a moving "dance floor" propelled by wind, tide, and climate.

Knowing what we know, and full of confidence that we understand waves both in theory and by actual test, we fling open the laboratory door, stride to the edge of the cliff, and look to the sea. Good grief! The real waves look and act nothing like the neat ones that endlessly roll down the wave channel or march across the blackboard in orderly equations. These waves are disheveled, irregular, and moving in many directions. No alignment can be seen between a series of crests; some of the crests actually turn into troughs while we are watching them. Should we slink back inside to our reliable equations and brood over the inconsistencies of nature? Never! Instead we must become outdoor wave researchers. And that means getting wet, salty, cold—and confused.

3   Wind Waves

There are many kinds of waves in the ocean, and they differ greatly in form, velocity, and origin. There are waves too long and low to see and waves that travel below the sea surface within water layers of different densities. Waves may be raised by ships, or landslides, the passage of the Moon and Sun, by earthquakes, or changes in atmospheric pressure. Probably there are kinds of waves that have not yet been discovered. But most waves, and the waves that are most important on a daily basis, are those raised by the wind.

Let us begin the life story of a wave with a perfectly smooth water surface such as a mirror-like pond. Suddenly a breeze begins to blow. Waves are born as the air pressure on the surface changes and the frictional drag of the moving air against the water creates *capillary waves* then *ripples*. When a ripple has formed, there is a steep side against which the wind can press directly. Now the energy can be transferred from air to water more effectively, and small waves grow rapidly. In the

PREVIOUS SPREAD: Wind waves are the creation of the atmosphere as it transfers its energy to the sea. Eighty-knot winds whip up waves in Drake Passage, Southern Ocean. *Colin Monteath/Minden Pictures/Nat Geo Image Collection* ocean the same thing happens, but with no nearby shore to limit wave development the waves soon develop into a sea.

Of course, it would be rare indeed for a constant wind to blow on an entirely undisturbed ocean surface. Usually there are "old seas," or waves generated earlier by winds elsewhere. If the new wind and these existing waves are moving in about the same direction, the old waves are rapidly enlarged. If the two are opposed, the wind will flatten the sea surface as the new waves cancel the old. For the moment, though, let us ignore this complexity.

Because winds are by nature turbulent and gusty, there are local variations in the air velocity and the pressure on the surface. As a result, wavelets of all sizes are created simultaneously.

As the wavelets continue to grow into larger waves, the surface confronting the wind becomes higher and steeper, and the process of wave building becomes more efficient—up to a point; there is a limit on how steep a wave can be. Steepness is the ratio of the height of a wave to its length, and the limit is about 1:7. A wave 7 feet (2.1 m) long can be no more than 1 foot (0.3 m) high. When small, steep waves exceed this limit, they break, forming whitecaps (a sea surface covered with such waves is said to be "choppy"). When the wind blows the top off a wave, causing a breaking wave at sea, some of the energy goes into turbulence but most is contributed to longer, more stable waves. The result is that a long wave can accept more energy and rise higher than a shorter wave passing under the same wind. Therefore, as the sea surface takes energy from the wind, the small waves give way to larger ones, which can store the energy better. But new ripples and small waves are continually being formed on the slopes of the existing larger waves. Thus, in the zone where the wind is moving faster than the waves, there is a wide spectrum of wave lengths. At the same time, however, the longest waves continue to accumulate energy from the smaller waves. Although the wind produces waves of many lengths, the shortest ones reach maximum height quickly and are destroyed, while the longer ones continue to grow.

Three factors influence the size of wind waves: (1) the wind velocity; (2) the duration of the time the wind blows; and (3) the extent of the open water across which it blows (otherwise known as the fetch).

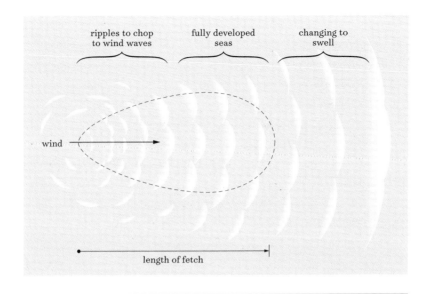

FIGURE 12: The fetch, inside the dashed line, is the area of water on which a wind blows to generate waves, transforming them from ripples to swell.

A simplified idea of the development of waves in the generating area is given in figure 12. In this generating area (often a storm), wind waves are called sea. At the upwind end of the fetch, the waves are small, but with distance they develop—their period and height increase and eventually they reach the maximum dimensions possible for the wind that is raising them. The sea, then, is said to be fully developed; the waves have absorbed as much energy as they can from wind of that velocity. An extension of the fetch or a lengthening of the time would not produce larger waves.

Drake Passage Crossing

In the austral spring of 2002, we left Punta Arenas, Chile, near the southern tip of South America. We headed toward the Southern Ocean, into the Drake Passage, bound for the Antarctic Peninsula. The experienced captain of the ice-strengthened ship said the winds were moderate—"only" 50 knots. We exited the Strait of Magellan, left sight of Tierra del Fuego, and headed farther south. The small, choppy wind waves of the channel were replaced by larger, long-period

rolling swells emanating from deeper water. Beyond the continental
shelf, the Southern Ocean's unlimited fetch produced waves over 35
feet (11 m) high. The decks were constantly awash. The winds and
waves increased, torturing the vessel, crew, and scientists for a few
days. I went to my bunk in high seas and when I awoke all was calm,
I was confused. The ship had ceased its twisting rolls and pitches.
The ship's engines were still running strong. I peered outside; we were
surrounded by mountains covered with glacial ice—we had entered
the Neumayer Channel. Protected from wind and deep-water waves,
it was calm again. A limited fetch and sheltered waters will brighten
your day. – KM

The description of how the wind transfers its energy to the waves
derives from the work of Harald Sverdrup and Walter Munk of the
Scripps Institution of Oceanography. During World War II their attention was attracted to the problem of predicting the waves and surf
that would exist on an enemy-held beach during amphibious landing
operations. *Wind, Sea, and Swell* gave the first reasonably quanti-
tative description of how waves are generated, become swell, and
move across the ocean to a distant shore. Today's changing climate
provides increased energy for transfer into ocean waves and allows
temperature, air pressure, and weather patterns to respond with
larger fluctuations.

SEA WAVES

Waves in a sea do not have the regular and precise properties of waves
generated in a wave channel. In a sea, the height of the crests and
the depths of the troughs are irregular, and the length of each crest
is short. These waves are individual hillocks of water with changing
shapes that move independently. The limits of a wave in a sea are
indefinable; each mass of water that the eye selects as a wave has a
different shape, a different speed, and a slightly different direction
from the other waves in the sea. The words *period*, *velocity*, and *wave*
length have lost the meaning they had in the orderly environment of
the wave channel. Try to determine the wave length of the waves in

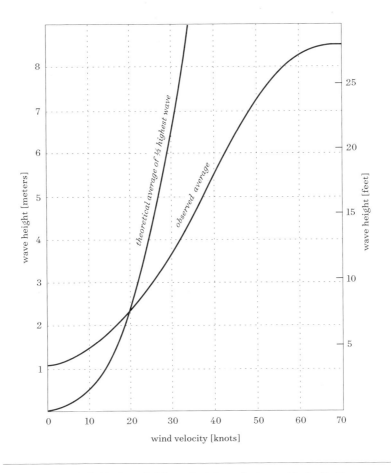

FIGURE 13: Wave heights and wind speeds observed in the North Atlantic (70,000 observations).

the photo on page 70. The spacing between waves is exceedingly irregular and demonstrates why statistical methods must be used to describe the properties of waves in a sea. Wave heights are nearly as irregular, but fortunately waves rise from a ready reference (the *mean sea level*) and there is somewhat less difficulty in defining wave height. For descriptive purposes, it is customary to use the average of the highest one-third of the waves (the significant wave height, $H_{1/3}$,

Under harsh conditions, waves can become so irregular and jumbled they are said to be "confused," with no recognizable height or length. Newhaven, East Sussex, United Kingdom. *Suerob/iStock*

FIGURE 14: The sum of many simple sine waves makes a sea.

see figure 13). The average of the highest one-tenth of the waves is sometimes noted as $H_{1/10}$.

Thus a sea is the result of superimposing a number of sinusoidal wave trains one on top of another, as shown in the accompanying conceptual diagram figure 14 above. Each layer represents a series of regular sine waves, as alike as those on a sheet of corrugated roof material, and has its own characteristic height, wave length, and direction. Individually the waves in these trains are as true to their classical formulas as those in the model tank.

The real instantaneous sea surface at any point is made up of all these layers added together. Where a number of crests coincide, there will be a high mound of water—but it will not last long, for the component waves soon go their own way. Similarly, a coincidence of troughs creates an unusually low spot, also of short duration.

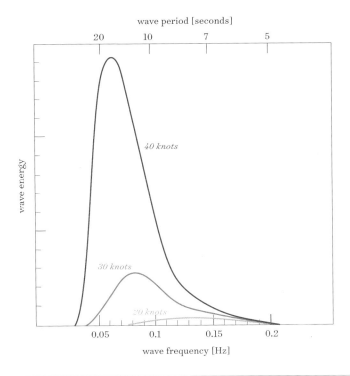

FIGURE 15: The wave spectrum for fully developed seas caused by winds of 20, 30, and 40 knots.

Since there are small wave crests in the large troughs and small depressions in the tops of the high mounds, on average the troughs and crests of the many layers of waves tend to cancel themselves out. The more layers of waves, the more random the sea surface and the lower the average wave height.

The need to reduce these complicated irregularities to a form that would be usable by the Navy in wave forecasting led scientists in the 1940s to a method of describing waves by means of their energy spectra. In this scheme a value is assigned to the square of the wave height for each frequency and direction. Then, after the portion of the spectrum where the energy is concentrated has been determined, it is possible to approximate average periods and lengths for use in wave forecasting. In other words, a wave spectrum gives a statistical description of wave energy and how it is distributed among various wave periods.

Some of the properties of wind waves are illustrated by figure 15, in which wave period is plotted against the amount of energy contained for three wind velocities. Each curve (or spectrum) represents the distribution of energy between various periods in a fully developed sea; the area under each curve represents the total energy. Consider first the 20-knot wind (a knot is a nautical mile per hour and thus equals 1.15 mph or 1.85 km/h). This relatively modest wind raises waves whose average height is 5 feet (1.5 m) and whose energy is spread over a band of periods ranging from seven to ten seconds.

If the wind increases to 30 knots, the waves become substantially higher and the periods longer. There is more energy available and these longer waves store it better. Now the average height is 13.6 feet (4.1 m), and the maximum energy is centered around a period of twelve seconds.

The upper curve of energy for the 40-knot wind waves shows a sharp peak at 16.2 seconds; the average height of the waves has increased to 28 feet (8.5 m).

Two things are clearly evident: As the wind velocity increases, (1) the amount of energy that can be stored by the waves increases greatly (this is because the waves are much higher and the energy is proportional to the square of the wave height), and (2) the periods become longer. (Note that in many scientific papers dealing with waves, the period, T, has been replaced by its inverse, frequency. Thus $f = 1/T$ and a ten-second wave has a frequency of 0.1 Hz.)

Table 3.1 gives the most important characteristics of seas that are fully developed for winds of various velocities. For example, a 20-knot wind must blow for at least ten hours along a minimum fetch length of 75 miles (120 km) to raise fully the waves it is capable of generating. When the sea from a 20-knot wind is fully developed, the average height of the highest 10 percent of the waves will be 10 feet (3 m). If a 50-knot wind were to blow for three days over a 1,500-mile (2,400-km) fetch, the highest tenth of the waves would average about 100 feet (30 m) high. Fortunately for ships, storms rarely reach such dimensions or durations, and we have refined our wave forecasting.

TABLE 3.1 Conditions in Fully Developed Seas

Wind	Distance	Time	Waves			
Velocity [knots]	Length of fetch [nautical miles]	[hours]	Average height [feet] (m)	H_3 significant height [feet] (m)	H_{10} Average of the highest 10% [feet] (m)	Period where most of the energy is concentrated [seconds]
10	10	2.4	0.9 (0.3)	1.4 (0.43)	1.8 (0.5)	4
15	34	6	2.5 (0.8)	3.5 (1.1)	5 (1.5)	6
20	75	10	5 (1.5)	8 (2.4)	10 (3.0)	8
25	160	16	9 (2.7)	14 (4.3)	18 (5.5)	10
30	280	23	14 (4.3)	22 (6.7)	28 (8.5)	12
40	710	42	28 (8.5)	44 (13.4)	57 (17.4)	16
50	1,420	69	48 (14.6)	78 (23.8)	99 (30.2)	20

Wave forecasting evolved during World War II and was refined throughout the world in the 1950s and 1960s. By the 1970s, the US Navy's Fleet Numerical Meteorology and Oceanography Center (FNMOC) had developed a spectral wave ocean model (SWOM). In the 1980s, the Helmholtz-Zentrum, in Geesthacht, Germany, developed another wave model called WAM (short for Wave Modeling). The British Oceanographic Data Centre has similar models and maintains an impressive amount of marine data, as does the Scripps Institution of Oceanography Coastal Data Information Program (CDIP), which provides coastal environmental data, wave models, and forecasting. Today, wave forecasting around the world is routinely used for optimizing military, cargo, and cruise ship routing. Proper routing reduces wave damage to vessels and fuel consumption. The ship captain, coastal engineer, beach researcher, and millions of surfers all rely upon the output from daily wave forecasting models. Kiss your wave forecaster.

Even in storms with lower-velocity winds there is always a statistical chance of a very high wave, called a rogue or extreme wave. No one can predict when or where or how high, but super-waves must exist because of the random nature of waves. For example, if 1,000 waves were observed on 20 different occasions, on one of those occasions the highest of the thousand waves will be 2.22 times the significant height. Thus, if the significant height were 44 feet (13.4 m), as it would be in a

The unfortunate bow of the aircraft carrier USS *Bennington* after encountering angry storm waves in a typhoon off Okinawa, Japan, in early 1945. The steel deck is 54 feet (16 m) above the waterline. *US Navy*

fully developed 40-knot sea, the "statistically exceptionally high wave" could be 97 feet (30 m) high.

Such a wave could exist only momentarily in a storm and it would be very unstable. It would tower over twice as high above most of its fellow waves, reaching upward into a mass of air moving at 40 knots. The crest would then be blown off, forming a breaking wave in deep water. It is these breaking storm waves—and they need not be super-waves necessarily—that do serious damage to the ships that are unlucky enough to be hit. The thousands of tons of violently moving water contained in the torn-off crest of even a moderate-size breaking ocean wave can destroy the superstructure of a ship.

The vast difference in the destructive power of breaking and non-breaking waves in deep water is worth examination, because it illuminates a fundamental property of waves. Objects in the water, such

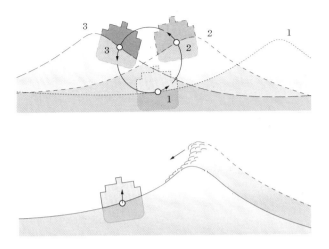

FIGURE 16: On top, ship and water particles in a large, nonbreaking wave have orbits of about the same size, so there is little motion relative to one another. On the bottom, water at the crest of a large wave has broken free of the orbit and will collide violently with the ship.

as ships, tend to make the same motion as the water they displace. A ship at sea in large waves will describe orbital circles that are roughly the same size as the water in that part of the wave. There is little relative motion between the bulk of the ship and the surrounding water. This motion of a ship may be uncomfortable, but it is safe (see figure 16).

If the crest breaks off a wave, however, the water moves faster than the waveform and independently of the orbiting water (and ship). The collision between the two could prove disastrous.

GREAT STORM WAVES

When sailors talk about the sea it is not long before they are on the subject of storms, great waves, and ship disasters. They speak of wave crests that are "mountainous" and troughs "like the Grand Canyon." However, when asked to assign dimensions to these features that can be used to test wave theory, their numbers are mostly guesswork—and likely to be on the high side to make their original story sound plausible. Because the heights assigned often do not seem to agree with theory, the question arises whether the eye has been deceived or the theory is inadequate. Moreover, the statistical explanation of wave variability makes it hard to say that any observation is wrong.

Newspaper accounts of exceptionally large waves encountered by ships on stormy passages are likely to relate to the deluge that occurs when the vessel drives her bow head-on into a wave. For example, if the water goes over the navigating bridge and the bridge is 100 feet above the waterline, a 100-foot (30-m) wave is reported. The unexpected impact of even a few tons of broken or "white" water at that level is no doubt a fearsome occurrence worthy of mention, but it is not evidence of wave height. Even if the water were part of a wave, the bridge is well forward of the ship's center, so that when the bow is down it is well below its proper level. The true wave height would be much less than 100 feet (30 m).

When visibility is good and a large ship is on a reasonably even keel, accurate estimates of wave height, even in a violent storm, are possible. The observer simply watches the distant horizon; when the crest of a wave obscures the horizon, that wave must be higher than the vertical distance between the observer's eye and the ship's waterline (see figure 17).

Stories of big waves at sea can be exciting. Vaughan Cornish, a British author who spent nearly half a century traveling the world on ships to collect data on waves, concluded that in North Atlantic storms, waves over 45 feet (14 m) high were fairly common; he reported several

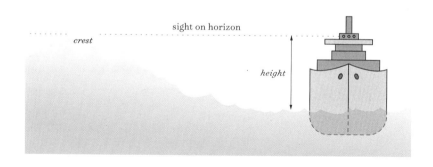

FIGURE 17: The crest of a wave will obscure the sight of the horizon when the wave is higher than the observer's eye. It can be a character-building experience.

well-authenticated examples of much larger ones. In his collection of data on wave length, he had many examples of storm waves 600 to 800 feet (180–240 m) from crest to crest, and swells two or three times that long.

How big can a wave in the ocean get? If one makes a set of assumptions about weather conditions that are extreme yet possible and plugs the data into a computer model, a (theoretically) large wave looms high above the heads of ocean design engineers. The European Commission research project MAXWAVE (2003) investigated rogue waves, using real data and models to estimate their effects on structures and ships, and to aid in future engineering designs. The following set of conditions would produce the maximum wave: (1) Time: The storm must be twelve to twenty-four hours old and the weather system must be dynamically stable over the same period; (2) Location: The geographic area where the storm occurs must be able to support a very low pressure zone. (The North Atlantic near Iceland, the Gulf of Alaska, and the Southern Ocean qualify.); (3) Distance: There must be a long, undisturbed fetch that allows large waves to develop. Pressure gradients of 900 to 980 millibars over a distance of 95 miles (150 km) must persist. This is a very steep gradient that can generate winds up to 150 knots. Computer model outputs show that a maximum combined height of over 200 feet (60 m) is theoretically possible.

ROGUE WAVES

Eventually some of the great waves predicted by statistical methods encounter a ship. They may sink it, but if there are survivors they will describe with awe the collision with a *rogue wave*. A rogue wave is a great solitary wave whose crest towers above the rest and scares the living daylights out of the luckless mariners in its path. Sometimes these waves express themselves as extra-deep troughs—"holes in the sea" into which a ship can fall, to then be overwhelmed by the next crest. It is a matter of debate which is more dangerous, but doubtless these super-crests and super-troughs account for many of the ships that have simply disappeared at sea with all hands. In the past twenty years, hundreds of container ships and supertankers, some over 660 feet

With decks awash, large waves torment a tanker in a storm. The contortions of the hull—creaks and shudders—are heard below the roaring of the winds by its trusting crew. On board the 1,140-foot (347-m) long *Esso Languedoc* off Durban, South Africa. The vertical masts are over 80 feet (25 m) tall. *ESA/Philippe Lijour*

(200 m) have sunk; rogue waves and heavy weather have sunk many more ships in the distant past.

A Cape Horn sailor, Captain William H. S. Jones, who made his observations from large sailing ships, once wrote: "It is strange but true that in high southern latitudes where seas 50 feet high and 2,000 feet long roll forward in endless procession, occasionally one sea of abnormal size will tower above the others, its approach visible for a considerable distance." It is easy to be impressed by a wave that towers over "ordinary" 50-foot-high waves.

In late 1942 the 81,000-ton, 1,000-foot-long (300 m) passenger liner *Queen Mary* was serving as a World War II troopship. On one occasion, while loaded with 15,000 American soldiers bound for Glasgow, she encountered a winter gale 700 miles off the coast of Scotland. The seas seemed very large, even to observers on a ship whose deck is some 60 feet (20 m) above the still-water level. Suddenly "one freak mountainous wave" struck the ship broadside. An eyewitness reported that the *Queen* "listed until her upper decks were awash and those who had sailed in her since she first took to sea were convinced she would never right herself." (Nearly all who have been to sea have been aboard a ship that rolled so far they were convinced she was going over.) But then, after hanging balanced on the brink of eternity for a few seconds that seemed very much longer, the *Queen* righted herself again.

On July 22, 1976, the tanker *Cretan Star* sailed from a Persian Gulf port loaded with 28,600 tons of light crude oil. On July 28, the master reported that the ship had encountered very heavy weather that had caused some damage and oil leakage. The last message received said, "vessel was struck by a huge wave that went over the deck and caused damage in the number 6 tanks—damage cannot be surveyed due to prevailing weather conditions." The ship was not far from Bombay at the time, and on August 2, searching aircraft from that port reported a black oil slick 4 miles (6.5 km) long and 1.5 miles (2.5 km) wide. That was all. A subsequent inquiry noted that when the southwest monsoon blows it reaches its greatest strength in July off Bombay and periodically piles up "episodic waves of vast proportions." The occurrence of rogue waves, sometimes called extreme waves (waves that are over twice the significant wave height) depends upon many factors including water depth and sea state. In *Waves in Oceanic and Coastal Waters* (2010) by Leo Holthuijsen, the statistics of deep-water and coastal waves are discussed in great detail. In a very general sense, one wave in roughly twenty-five is over twice the height of the average wave, one in 1,000 is over three times average height, and only one in 300,000 exceeds four times the average height. Although infrequent, if you encounter a rogue wave at sea, you will remember.

One part of the ocean is so well known for rogue waves that all mariners were warned in advance by the British Admiralty's book of sailing directions known as the *Africa Pilot*, published in 1878, and still in print. This danger zone is off the "wild coast" of South Africa between Durban and East London. A few miles offshore the gentle slope of the continental shelf reaches 100 fathoms (180 m) and then drops precipitously away into deep water, thus creating a wall-like barrier against which the Agulhas Current presses hard. This massive flow of water moves southwest at high velocity—often at 4 knots (1.2 m/sec) and sometimes 6 knots. The strongest current is in the main shipping lane moving south—a lane chosen by the ship operators themselves.

When a gale blows from the southwest and raises waves that move directly against this current, the wave length is shortened and the wave steepness greatly increases. Even when there is no local storm, a large swell coming north from the Antarctic Ocean meets the south-flowing current head-on and creates impressive waves called "Cape Rollers" or "Agulhas Swell." Under some circumstances the unusually swift current actually doubles the height of the waves.

The question is: *why*? There are several explanations. One reason is that approaching waves are refracted, or bent, toward the higher current velocity, especially at an abrupt change in velocity such as occurs along a steep continental boundary. This concentrates the wave energy over the strongest current and focuses certain waves there. In the special case of a ship moving with the current, the ship's velocity causes an additional steepening of the wave (relative to the ship). Moreover, if the ship is moving at roughly 18 knots (9 m/sec) in one direction aided by a current of 8 knots (4 m/sec) and encounters an oncoming wave moving through the ocean at 19 knots (10 m/sec), the velocity of the collision is the sum of these, or 45 knots (23 m/sec). Because the force of impact is proportional to the square of the velocity, in this very special case the current nearly doubles that force. If the wave is twice as high as an ordinary storm wave, a ship is likely to be in trouble.

This phenomenon has long been the subject of scientific attention, and it was known for many years before Professor J. K. Mallory of the University of Cape Town called attention to the concentrated loss of ships. Early Portuguese caravels were lost in this area. The losses have continued; for example, the passenger liner SS *Waratah* with 211 persons aboard that vanished there in 1909. (Off the Cape Hatteras coast of

Smaller vessels are too small to span between wave crests. The larger waves heave and rotate the entire vessel like a cork. Grand Banks of Newfoundland. *Philip Stephen/NPL/Minden Pictures*

the United States, jets of the Gulf Stream seem to have caused the loss of ships in a similar manner.)

In 1973, the *Ben Cruachan*, a new British cargo vessel of 12,000 tons, was moving southwest off the coast of Durban, South Africa, when suddenly the bow of the ship was seen to extend out over a hole in the sea. A large, closely following crest then smashed down on the bow with such force that it broke the ship's back. Similar incidents with equally bad results have been reported by the ships *Neptune Sapphire* (12,000 tons), *World Glory* (45,000 tons), and *Wafra* (70,000 tons). The supertanker *World Horizon* (102,000 tons) broke up in heavy swell off of South Africa, and ships of more than 200,000 tons have also been sunk. Rogue wave sinkings are not mainstream news.

Ships traveling this part of the South African coast, offshore another 20 miles (32 km) or so, would be well outside the part of the current that caused the damaging waves. However, the temptation of a free ride on a current of 4 knots or more, with its substantial savings in time and fuel, is too great. Generally, the largest ships are the hardest hit. Because of its great mass, a supertanker responds to waves more like a seawall than a floating object and tends to resist rigidly instead of giving way as a smaller ship would do. Occasionally, though, a small vessel encountering a rogue wave makes it into the news. Sebastian Junger's book *The Perfect Storm* immortalized the loss of the fishing boat *Andrea Gail* (only 92 tons) off the coast of Nova Scotia in October of 1991.

INTERNAL WAVES

The opening sentences of Chapter One described waves as undulating forms that can travel on the interface between any two fluids of different density. The surface of the sea is the obvious place to study waves; however, other waves that are quite different in character travel on the interface between layers of slightly different densities within the ocean. These are *internal waves*. Imagine pulling a parcel of water down into more dense water, then letting it go; it will rise because it is less dense than its surrounding water. But when it has risen enough to be neutrally buoyant, it continues to move as its inertia causes it to rise into less-dense water. When its upward motion stops, it sinks back downward. This repeating process is the essence of internal wave motions. It is similar to a hanging weight oscillating on a spring where the buoyant force is represented by the spring. Because waves are driven by gravity, it follows that if there is a substantial difference between the densities, as there is between air and water, the forces that tend to restore a flat surface are great. If the density difference is small, as it is between layers within the ocean, the restoring forces are much weaker. The period at which an internal wave will oscillate is related to the water's density gradient (stratification) and is called the Brunt–Väisälä frequency, or buoyancy frequency, N. N equals the square root of gravity divided by the water density of the parcel multiplied with the change in water density divided by the change in depth.

$$N = \sqrt{ \frac{\text{gravity}}{\text{water density}} \times \frac{\text{change in water density}}{\text{change in depth}} }$$

A larger value of N indicates a more stable (stratified) water column. This is important in the real world when less-dense river water flows into the ocean and when glaciers and sea ice melt. Waves are always working, mixing nutrients and dispersing pollutants, even while massaging sediments below. All are affected by gravity, buoyancy, and ocean stratification.

The wave periods, lengths, and heights of internal waves can be very much greater than those of surface waves. Although the heights may be large—over 330 feet (100 m or more)—the energy these waves contain is smaller than surface waves and the typical velocity is low, averaging less than 1 foot per second (30 cm/sec).

It is not essential to have an abrupt density interface; any stable density stratification can support internal waves. And because there is usually a relatively warm surface layer over much of the ocean, internal waves are a common phenomenon. Although the major part of these waves is well below the surface, evidence that they exist can often be seen, especially on calm, clear days. This is because internal wave currents affect the reflectivity of the sea surface by producing alternating bands of slicks and (small-scale) roughness.

Internal waves are important in several ocean processes. For example, the cold, dense water formed in polar regions sinks and spreads over the bottom of the ocean basins, and somehow it must be mixed with the waters above if a reasonably steady state is to be achieved. Internal waves are a significant cause of this mixing, and they may play an even greater role in the transfer of momentum, especially between layers of water that are moving in different directions. The intermittent mixing and uplift of phytoplankton (tiny plants that are the foundation of the oceanic food web) into the sunlit surface water by the passage of internal wave crests leads to increased biological productivity. The power continually dissipated by internal waves is immense and has been calculated to be as large as the power used by humans on Earth; some estimates are over 2 terawatts continuously, for all 8,736 hours per year (or 17,000 terawatt hours per year).

Internal waves are generated by the addition of downward energy from external sources. In principle, they can be set in motion by moving atmospheric pressure fields (i.e., weather fronts), variable wind stresses, surface waves, tides, ships, and downward-moving or upward-moving currents.

Sensing the presence of these huge, slow-moving waves requires a great many measurements. Researchers from the Scripps Institution of Oceanography recorded shoreward-moving bands of variable roughness by means of a time-lapse camera on a cliff top. A series of temperature versus depth measurements using a towed thermistor (a type of temperature sensor) chain hundreds of feet deep (200 meters deep with a temperature sensor every meter, recording on shipboard) was used extensively by Woods Hole oceanographers in Massachusetts to obtain a reasonably detailed picture of internal waves as revealed by temperature variations. Internal waves are observed almost everywhere when there are fluid layers with different densities.

The Internal Wave Mystery

Internal waves can still confuse researchers. In the early 1990s, I was conducting some autonomous vehicle research in the deep waters off of Bermuda. A vehicle was deployed and programmed to descend to 3,000 feet (1,000 meters), taking sensor measurements along the way, before becoming neutrally buoyant at 3,000 feet (1,000 meters), and then returning to the surface. The vehicle dynamics were well understood, and we could comfortably predict the vehicle mission times with great accuracy. However, our first deployment off Bermuda was a "knuckle-biter" because the time required for the vehicle to return to the surface strangely doubled. We inspected the vehicle for defects, found none, and deployed it for a second time. Again, the vehicle misbehaved. Weeks later, after we examined the data, we solved the mystery. We had not anticipated the vertical movement of the large 330 foot (100 m) internal waves. Our neutrally buoyant vehicle had been "riding" the immense internal waves! We observed similar events in the Pacific with other vehicles. Today, this neutrally buoyant technique is exploited to understand the motion of internal waves and their influence on biology and ocean mixing. All parts of the ocean have their "ups and downs." – KM

High-frequency underwater sounds make marvelous picture-like acoustic records of internal waves. This is possible because "pings" of sound hundreds of kilohertz from instruments are backscattered by the billions of tiny animals called zooplankton that are concentrated in layers of constant density or temperature. The returning echoes from depths to about 150 feet (50 m) are recorded and, ping by ping, a graphic representation develops of the changing pattern as internal waves pass. A specialized system can detect water turbulence that does not have abundant animal life. An objective of this type of work is to understand ocean mixing and biological productivity. When the tide in summertime comes into Massachusetts Bay, for example, it generates a packet of internal waves as it passes Stellwagen Bank. Sometimes the surface influence of these waves (alternate bands of slick and rough water) can be seen by radar. Over the years, radar has become increasingly useful in detecting the properties and the wave motions of the surface of the ocean. Radar can use radio (electromagnetic) signals from a few MHz to over 100 GHz depending on the phenomena to be observed.

OIL, WAVES, AND ICE ON TROUBLED WATERS

The calming effect of oil on the sea surface has been known for many centuries, long before the physics of the action was understood. All oils are not equally effective; experience has shown that fish oils or other viscous animal oils are best and that petroleum products have relatively little effect. Because the latter are much more likely to be available around a modern boat, there have been many attempts to use motor oils, without success. As a result, the idea of using oil to calm the sea surface has fallen into disrepute, to be regarded as an old seafarer's tale without foundation in physics. But properly used, oil can be very helpful to the small boat operator under emergency conditions. However, it is "leaked" on rather than "poured" on the rough waters.

There is no doubt whatsoever that this method works, but one must not expect too much. A thin film of oil could hardly be expected to have any effect on large waves or swell, but it does quickly extinguish the small waves. Moreover, as the sea surface becomes slick, the wind has less effect on it, no spray is blown about, and the wave crests become more rounded.

Kelp beds protect the shoreline by smoothing out the shorter-period waves. This process creates the "glassy" areas inside the kelp. Leo Carrillo, California. *Kyle Sparks*

This smoothing is caused by increasing the *surface tension* of the nearby area of sea. The higher the surface tension of the liquid, the stronger this invisible membrane acts. But because engineering handbooks give a surface tension for water twice that of oil, it is not readily apparent how the addition of oil can help matters in times of distress.

One answer is that the surface tension of the oil increases as its thickness decreases. The thinner the film the better, for oil can act like an elastic membrane even when it is only a millionth of a millimeter thick. Thus, as the oil spreads away from the boat it becomes more effective, opposing any motion that tends to increase the surface area. At a distance from the boat, depending on the velocity of the wind, the elastic limit of the increasingly thin film is exceeded. It breaks up and blows away, making it necessary to continually add more oil at the center. Professor and mariner Charles "Chip" Cox published a paper in 2017 about this interesting yet poorly understood phenomena.

Other materials that are mixed in the water or are floating on it also tend to reduce wave action. Extremely muddy water, for example, will cause waves to decay rapidly, and so will masses of floating debris. Breakwaters have been constructed with buoyant subsurface objects tethered to the seafloor. Kelp (large brown algae) grows just offshore in the temperate seas of the world. The large waves coming from afar pass through the kelp beds almost unchanged, but the small waves caused by local winds are quickly dissipated, and the water surface is nearly always glassy just inside the kelp beds. Globally, kelp partially shields about a quarter of the Earth's coastline from waves. Some areas have experienced a decrease in kelp bed coverage due to increasing seawater temperatures.

Lost in the Fog

Divers frequently consider wave height, direction, tidal currents, and visibility when planning a dive. I was freediving with a buddy in the kelp beds off Northern California; fish were plentiful, and all was going well. The kelp undulated with the peaks and troughs of the large swell as it passed through the kelp bed. The strengthening tidal current aligned the kelp with its flow. From the crest of each wave I could briefly view the coastline. Then the fog rolled in and we lost sight of land. After a few more dives I carelessly lost sight of my dive-buddy and drifted with the current out of the kelp. The thick fog had obscured the sun and reflected waves from afar made the swell complex. Soon I had no sense of direction. I was moving with the chaotic orbital wave motions and drifting at sea. The harrowing experience continued until a break in the fog allowed me to reorient myself and find my dive-buddy back in the kelp. He knew I was gone but was unaware that I had ever been adrift, confused by the swell and without any point of reference. A point of reference is good in a chaotic world. – KM

Fields of sea ice also reduce wave action; as waves move through the ice pack, the shorter ones are "damped out," and there is an apparent increase in wave length. Ice, therefore, serves as a protective shield. Any ice cover reduces the amount of wind energy the ocean is willing to accept and reduces the undulating forces of waves upon the shores.

The marginal ice zone, where the ice is churned by the waves, is a floating battleground. All the short-period waves die first, leaving behind interesting forms of what is called "pancake ice," the remnant of the mixing of wave energy with the slush of ice crystals. Only the strongest and longest-period ocean waves survive to propagate into multiyear icefields. Long-period waves have been measured more than 60 miles (100 km) into polar ice. The cartographer Robert Perry produced one of the first detailed charts of the Arctic in 1985. It revealed many unknown features across the Greenland Abyssal Plain, Fram Strait, Yermak Plateau, Belgica Bank, Kane Basin, Beaufort Sea, and Northwind Ridge. Although submerged below, these bathymetric shapes determine both the propagation of waves and the circulation of the Arctic Ocean.

Waves in the Ice

In the 1980s, when sea ice coverage was much heavier, I was involved in many expeditions to the Arctic. Some of these expeditions were in the marginal ice zone off the northern coast of Greenland. Our research projects included icebreaker, helicopter, and ice camp operations during which we frequently encountered polar bears on the ice. An icebreaker served as our "mothership." From the icebreaker we would venture out, walk, snowmobile, and helicopter to remote locations to collect wave, ice, and oceanographic data. When we were on the ice, away from ships and machinery, the subtle ebb and flow of energy became noticeable.

On one occasion, two of us ventured forth alone several miles from the icebreaker. The icebreaker had already been "made fast" to the thick multiyear ice with "ice anchors." The main ship's engines were shut down, and quietness reigned as we stepped on the ice. When we first stopped the snowmobile perhaps a half mile (1 km) from the icebreaker, the sound of the wind, our footsteps, and our breathing were all that we could hear. We stopped a second time, about 3 miles (5 km) from the

PANCAKE ICE Pancake ice is formed by waves passing through ice. Random wave motions cause the ice to bash into its neighbors and round off any square edges. Weddell Sea, Antarctica. *Anne-Mari Luhtanen/Finnish Environment Institute SKYE*

breaker to drill through the ice and take some measurements. Whenever the squeaking, crushing noises from our footsteps ceased, there was almost silence, but not complete. A more subtle intermittent creaking sound began to emerge.

We continued to work on the frozen sea, isolated from the world, a thousand miles from any inhabited land. As we began to tire, our periods of rest increased. It was 20 degrees below zero. We listened more, and detected a rhythm to the creaking. We came to understand that it was the long-period swell fracturing the multiyear ice with its wave energy. We had not noticed the tiny fractures beneath us until the wind direction changed.

Larger fractures were appearing when suddenly the creaking sounds were replaced by the sound of a cannon. In the distance the shifting winds had caused one of the ship's ice anchors to pull loose. The stretched mooring line connected to the anchor had stored enough energy to catapult a 400-pound (180-kg) anchor crashing into the bridge of the icebreaker. We watched as the 400-foot (122-m) Polar-class icebreaker slowly drifted away, blown by the wind. Stronger winds further relaxed the ice, and the fractures became larger gaps of exposed sea. The long-period swell had won.

We were adrift, marooned; floating on an ice floe. The liquid sea expelled a mist of sea smoke, which obscured our views. Amid the pandemonium aboard the icebreaker and the efforts to restart the main engines, we had been forgotten. Alone, we imagined polar bears in the mist and certain death. Eventually, the crew of the icebreaker refocused their attention to their drifting comrades and ended our chilling experience. Unseen waves are everywhere you look. – KM

SWELL

As waves move out from under the winds that generated them, their character changes. The original wind waves are said to decay, meaning the crests become lower, more rounded, and more symmetrical. They move in groups of similar period and height, and their form approaches that of a true sine curve. Such waves are now called *swell*, and in this form they can travel for thousands of miles across deep water with little loss of energy.

In more formal language, these waves are "periodic disturbances of the sea surface under the control of gravity and inertia and of such height and period as to break on a sloping shoreline."

The usual range of period of swell is from six to sixteen seconds, but occasionally longer periods are clocked. The average period of the swell arriving at the US Pacific coast is slightly longer than measured in the Atlantic. This difference arises partly from the much greater size of the Pacific, in which longer waves can be generated in larger storm areas, and partly from the greater distances the waves must travel across the Atlantic continental shelf before they reach the shore—in which the longer-period waves are attenuated.

As swell, the waves have relationships conforming rather closely to the simple equations that applied in the wave channel. That is, their wave length L is about $5.12\ T^2$, where T is the wave period in seconds (or in meters $L = 1.56\ T^2$), and they move at a speed of $\sqrt{\dfrac{gL}{2\pi}}$.

TABLE 3.2 Approximate Lengths and Velocities of Sinusoidal Swell in Deep Water

T Period [seconds]	$L = 5.12\ T^2$ Wave length [feet] (m)	Velocity [feet per second]	C Approx. velocity [mph]	$d/L=0.5$ Water depth [feet]
6	184 (56)	30.5	21	92
8	326 (99)	40.6	28	163
10	512 (156)	51.0	35	256
12	738 (225)	61.0	42	369
14	1,000 (305)	71.5	49	500
16	1,310 (399)	82.0	56	655

Table 3.2 gives the range of wave lengths and velocities that swell in deep water would have if it were truly a sequence of regular sine waves. It is not perfect, but this approximation is adequate for now. The longest swells reported have periods of more than twenty seconds; these waves have wave lengths over 2,500 feet (760 m) and a velocity of 70 mph (110 km/h).

Swell moves across the open ocean between the generating area and the distant shore in trains made up of groups of waves. These trains are

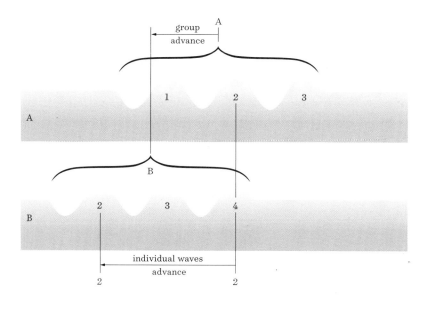

FIGURE 18: As a group of waves advances, the first wave (shown in A) uses up its energy and dies out as a new wave (shown in B) forms behind. The center waves continue at normal velocity until they, too, die out. This results in the *group velocity* of deep-water waves being only half that of an individual wave.

bundles of energy. Although any wave in the group moves at a velocity that corresponds to its length as indicated above, the velocity of the group as a whole is only about half as fast. As the wave train moves into an undisturbed area, it must use some of its forward-moving energy to set new water particles orbiting; this energy is contributed by the leading waves, with the result that these constantly disappear from the front of the advancing wave train. Because new waves constantly form at the rear of the train, the number of waves and the total energy remain about the same, but the velocity of the group as a whole is reduced (see figure 18).

It is easy to see that the understanding of *group velocity* is most important to those who forecast the arrival of waves from distant storms. A wave group whose period averages twelve seconds will take two days to cross 1,000 miles (1,600 km) of open ocean instead of half that long, the speed of an individual twelve-second wave.

This phenomenon was first reported by John Scott Russell in 1844, and it can be observed in a long wave channel if the wave-making machine is operated for only a few strokes. The observer, walking alongside the tank and focusing attention on the first wave generated, will see it decrease in height and finally disappear altogether into the undisturbed water ahead of the group. The last wave in the group is usually not so clearly defined as the first, but sometimes a wave can be seen to develop behind the last wave generated.

Thus the composition of a group of waves is constantly changing and the individual waves observed at a shore are but remote descendants of those actually generated by a distant storm.

The energy possessed by a wave is twofold in nature. In part it is kinetic energy, due to the motion of the water particles in their orbits; the remainder is potential energy due to the elevation of the mass of water's center of gravity in the crest above sea level. For swell, averaged over several periods, the two energies are equal.

Eventually the deep-sea swell approaching a coast moves into the shallow water of the continental shelf, and when the depth of water is less than half the wave length, these waves feel the bottom and undergo some radical changes in length, velocity, and direction.

4 Waves in Shallow Water

We have defined shallow-water waves as those that are traveling in water whose depth is less than half the wave length. Thus, whether a wave is in shallow water depends on the basin as well as on the wave. In a wave channel with water 2.5 feet deep (75 cm), waves with periods longer than one second are shallow-water waves; in 600 feet (200 m) of water at the edge of the continental shelf, a sixteen-second wave is in shallow water, and in the deep ocean basin, where the average depth is about 15,000 feet (4,500 m), all waves with periods greater than eighty seconds are considered shallow-water waves.

Therefore, this chapter deals with the entire range of wave periods as illustrated in figure 19. On the continental shelf, large ocean swells move predominantly as shallow-water waves, and in the deep ocean, tsunamis and tides are very-shallow-water waves. As these various waves approach shore and move across shallow water they react in special ways. They *reflect* (turn back by a vertical obstacle), *diffract* (spread), and *refract* (bend around) their energy, which means that they

PREVIOUS SPREAD: Shallow water influences all waves as they swell, compress, and then break on the shore. Banks Avenue, Mount Maunganui, New Zealand. *Rambo Estrada*

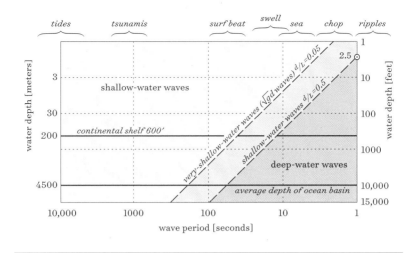

FIGURE 19: Shallow-water and deep-water waves.

work their way into the water behind projecting rocky promontories, and adapt to a gradually shoaling bottom. The great low waves of the deep sea may move up an estuary (where the salty water intermingles with fresh water) with an abrupt steep-front, or they might break down into a dozen smaller waves. On entering a bay or harbor, they might excite it in such a way as to cause a sloshing motion. In short, shallow-water waves have just as complicated characteristics as deep-water waves, and they are likely to be more interesting, because it is in shallow water that they most affect humankind.

Reflection

When a wave encounters a vertical wall, such as a steep rocky cliff rising from deep water or the vertical end of a wave tank, it reflects back on itself with little loss of energy. If the wave train is regular in period, a pattern of standing waves may be set up in which the orbits of the waves approaching the cliff and those reflected by it modify each other in such a way that there is only vertical water motion against the cliff and only horizontal motion at a distance out of one-quarter wave length, much as shown in figure 20 on page 100. Regular wave trains are rare in nature, and this unusual circumstance is called *clapotis*. The point is that as long as the wave is roughly sinusoidal, it exerts relatively little force on the structure that reflects it. Therefore, when possible, breakwaters are constructed in water too deep to cause waves to break. The forces that would be imposed on the same structure by breaking waves are far greater, as we will see presently.

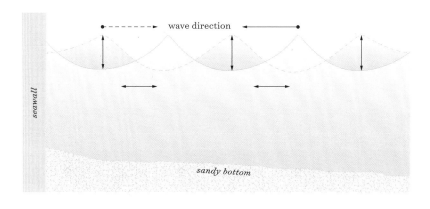

FIGURE 20: Wave reflection (clapotis) from a nearly vertical wall may result in standing wave patterns in the water.

Virtually any obstacle will reflect some part of the wave energy. An underwater barrier such as a submerged coral reef will cause reflections, even though the main waves seem to pass over it without much change. As mentioned earlier, the Polynesians used such propagating wave patterns to navigate great distances. If the light is right, the reflections can be seen from the air as a halo of small waves surrounding the reef, bucking the main swell. Or, a steep beach may reflect waves to a remarkable extent—and as the reflected waves move outward and encounter the incoming waves head-on, thin sheets of water may shoot upward 20 feet (6 m) or more. If the two meet at a slight angle, a "zippering" effect is observed as the point of impact races along at as much as 100 feet per second (30 m/sec). I have observed the waves of translation (or foamlines) from large breakers strike steep cobble beaches and return seaward as reflected waves 6 feet (2 m) high. When transitory waves of this size collide there is a roar, and much water is thrown about in confusion.

WAVE REFLECTION A reflected wave moves back seaward against an incoming wave, and the motions of the colliding waves (kinetic energy) thrusts the water upward. The encounter started at the right and is moving to the left. Santa Barbara, California. *Mike Eliason/Santa Barbara County Fire Department*

DIFFRACTION

Imagine that a train of waves (i.e., swell) moving across the ocean suddenly encounters a steep-sided island rising abruptly from the depths. Anyone would expect the waves to be lower on the lee side, and a boat seeking calmer water would run in behind the island. But exactly where would the boat stop rolling as it moved into more protected water? Would the island cast a clear "wave shadow" in which the water is perfectly calm? The answer is no; the reason for this is that waves diffract. As the waves pass the island, some of their energy is propagated sidewise as the wave crest extends itself into the area apparently sheltered by the island. This is diffraction. It occurs when waves spread out their energy as they pass around the edge of an obstacle or through any opening.

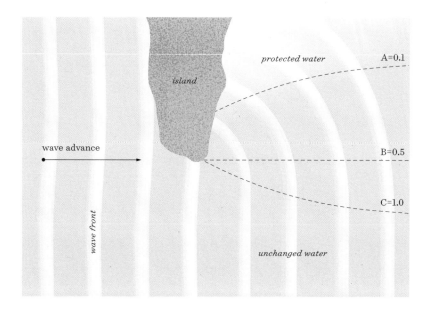

FIGURE 21: Waves passing into an island's lee will spread out in a circular arc—this is called diffraction. The wave height will decrease exponentially as it moves into undisturbed waters. The wave energy originally concentrated between C and B is spread between C and A.

Figure 21 shows this process. When a train of regular waves passes an island whose geometric shadow is indicated by the dashed line B, some of the wave energy from the region between B and C flows along the crest into the region between B and A. The numbers given are approximate diffraction coefficients for this hypothetical case. That is, at C the waves are full height, at B they are only half the original height, and at A they are one-tenth the height of C.

The phenomena of wave diffraction must be taken into account by engineers who design breakwaters. Otherwise, ships moored inside, apparently in a safe lee, might be damaged by wave action. In addition, diffraction need not be a shallow-water effect. This example specifies a steep-sided island rising abruptly from the depths. If the same waves had approached the island over a gently sloping underwater topography, the result would have been entirely different because of wave refraction.

WAVE REFRACTION Wave refraction is seen at this headland. The waves bend as they pass into shallow water, adjusting to be parallel to underwater contours and thus the coast. Sandon Point, New South Wales, Australia. *Andrew McInnes/Alamy Stock Photo*

REFRACTION

Refraction simply means bending. As waves move into shoaling water, the friction of the bottom causes them to slow down, and those in the shallowest water move the slowest. Because different segments of the wave front are traveling in different depths of water, the crests bend, and wave direction constantly changes. Thus, the wave fronts tend to become roughly parallel to the underwater contours.

A simple example of refraction is that of a set of regular waves approaching a straight shoreline at an angle. The inshore part of each wave is moving in shallower water, and consequently moves slower than the part in deep water. The result is that the wave fronts tend to become parallel to the shoreline. Thus, an observer on the beach always sees the larger waves coming in directly, even though some distance out from

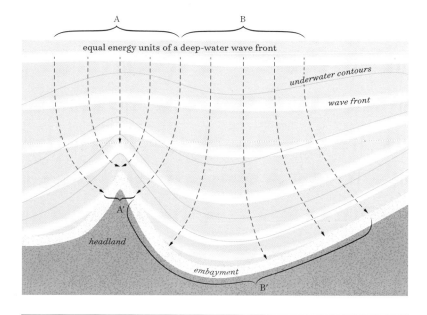

A   B

equal energy units of a deep-water wave front

underwater contours

wave front

A′

headland

embayment

B′

FIGURE 22: Wave refraction causes a straight wave front in deep water to bend until it almost parallels the shoreline. Equal energy units in deep water (A and B) are concentrated on the headland or spread along the embayment (headland A′ and embayment B′).

shore they are seen to be approaching at an angle. The effect of refraction is to concentrate wave energy against headlands as shown in figure 22. It is a process of considerable geological significance, a modern expression of the old sailor's saying: "The points draw the waves." (For more information on this, see figure 6 on page 38 in Chapter One, which shows the result of wave refraction acting over a long period of time.)

Wave Refraction into Smugglers Cove

In April of 2015, I was provided with a character-building lesson on wave refraction. The plan was simple: I would be dropped off on the windward side of Santa Cruz Island and was to be met by a sailboat on the leeward side. Although Santa Cruz Island is the largest of the islands off the coast of Southern California, it has no permanent residents. Large waves propagate into the Santa Barbara Channel from the northwest and dissipate their energy by breaking along the northern shore or refracting around the eastern end of the island. When the waves are just right, sea caves resonate with the wave's energy, spew spray back into the air, and claim victory over the waves. The Native American Chumash have successfully navigated this perilous channel for thousands of years. And while my hike from Scorpion Anchorage to Smugglers Cove lasted only a few hours, getting from the beach to the awaiting sailboat offshore was an adventure.

By the time I approached my destination, the winds on the northern side of the island had increased and the wave height had built immensely. As I reached the lee of the island at Smugglers Cove, strong katabatic (density-driven) winds were blowing offshore. The 60-knot winds battled with the waves, swirling across the deserted, windswept shore. The waves, elevated by the winds and bent by the bathymetry, refracted around the island. Their energy filled the cove and broke on the shore, and I was marooned on the uninhabited portion of the island. There was no fresh water, no food, no cell phone coverage, or any remedy for my underestimation of wave refraction. After a long, sleepless, near-hypothermic night, I welcomed the dawn. By sunrise, the winds had diminished, yet the refracting waves persisted, barring the possibility of any vessel landing on the south side of the island. Several dead marine mammals were on the beach, and I thought of great white sharks, which are common in the area, as I reluctantly stepped into the 54 degree (12° C) water. Each wave sucked more heat from my body as I swam through the surf into deeper water and the awaiting dinghy. My lesson—do not underestimate wave refraction; it will propagate into every corner of the universe, including Smugglers Cove. – KM

WAVES IN SHALLOW WATER

Even on a *circular island* (with a gently sloping underwater topography on all sides) a wave train from one direction will wrap itself around the island so that the wave fronts arrive nearly parallel to the beaches on all sides, although of course the waves are substantially higher on the side facing directly into the deep-sea swell (see figure 23, page 106).

As a train of swell becomes a shallow-water wave group where the depth is less than half the wave length, the effects of refraction begin.

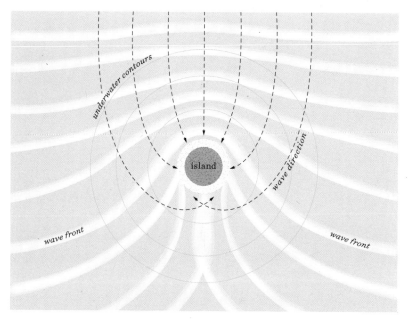

FIGURE 23: Waves refract around a hypothetical circular island. Waves are moving down the page.

One can compute the velocity of a wave in shallow water, and it is possible to make a diagram showing how far the various parts of a wave advance during a series of equal time intervals. Such refraction diagrams are very useful in visualizing how waves of various periods and directions will influence a shoreline or a proposed coastal structure.

There are several methods of drawing refraction diagrams; all begin with an accurate contour chart of the bottom configuration (bathymetry) out to a depth of half the longest wave that will be considered. Then, the period and direction of the waves to be diagrammed must be selected. The practicing coastal engineer will prepare diagrams for waves of many periods and directions, and a statistical wave "hindcast" will also usually be made. That is, estimates of wave heights and periods will be made based on historic weather maps to obtain statistics on what waves have arrived in the past and are likely to arrive in the future. However, this becomes more difficult as our climate continues to change. One of these waves is likely to be predominant and will serve as a good start—on much of the US West Coast it is often a twelve-second wave from the northwest. The engineer proceeds by drawing a straight line representing a wave front in deep water, or using another method, a wave ray (perpendicular to the wave front) that shows the direction of wave advance. In the wave-front method it is customary to calculate the new, somewhat reduced wave length for each contour depth and to use these to step off the advance of the wave front. The resulting diagram shows the successive positions of the wave front at time intervals equal to the period. As the wave slows down, the wave fronts get closer and closer together.

The principal question answered by a refraction diagram is: *How is the wave energy distributed when it reaches the coast?* If the deep-water wave front is divided into equal parts and wave rays are drawn through these perpendicular to each wave front, the energy distribution is readily seen. These wave rays separate areas of equal wave energy and are called *orthogonals*. The ratio of the length of the wave crest between orthogonals in deep water to that at the beach is the *refraction coefficient*. With it, one can compare the amounts of wave energy reaching various points along the coast and determine the effectiveness of proposed breakwaters.

This method is universally used to help predict future conditions as well as using the data to hindcast the conditions that have caused damage in the past.

TRAPPED WAVES

One effect that refraction can have where the near-shore undersea topography slopes rather steeply toward deep water is trapping waves that strike the beach at an angle. As shown in figure 24 on page 108, which illustrates a wave front approaching a beach at an angle, the first part of the wave front to strike the beach is reflected and already moving seaward when the next part of the front reaches the sand. The figure shows successive positions of a single wave at intervals of about five seconds.

Trapping is evident in the dashed orthogonal line; this line shows the direction of motion of the wave energy and thus is always perpendicular to the wave front. The bundle of energy (in red) moving with this wave reaches the beach at point B; it is reflected and we see the orthogonal

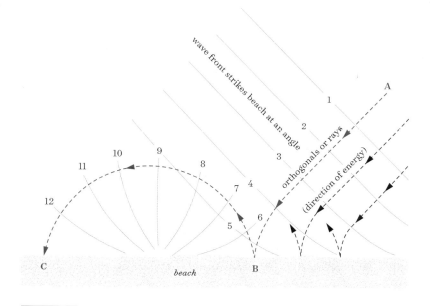

FIGURE 24: When waves strike the beach at an angle, part of the energy that is reflected is "trapped" by refraction and moves along the beach. Numbers indicate successive positions of a single wave front.

start out to sea but then bend around and eventually strike the beach again at C. A great deal of the energy is lost when the wave breaks and swashes up the sand, but, depending on the circumstances, up to about 30 percent of the energy could be reflected to continue on. At the outer edge of the reflected wave (stages 7 and 8), the wave front stretches and more of the energy radiates off into deep water by diffraction. But much of it follows the orthogonal so that the wave at stage 12 is similar to that at stage 5, but much smaller.

Trapped waves were discovered not by observation but by studying the implications of refraction; the first description of them was given by Professors John Isaacs and Carl Eckart of the Scripps Institution of Oceanography in 1952. Later, Dr. Walter Munk, also of Scripps, provided a more detailed mathematical treatment of what he called edge waves.

Because a reflected wave crest sometimes returns to sea as several smaller waves of shorter period, it is not quite clear how one should draw the refraction diagrams for these waves that move edgewise along

CUSPS Cusps are crescent-shaped repeating forms on the beach face. These features are cut out in the center and taper seaward on both sides like stubby fingers (for more see Chapter Ten). Dorset, United Kingdom. *Dr. Stanislav Shmelev*

the coast; figure 24 is intended only to show the principle of trapping. It may be that this process is most important in the alongshore propagation of the energy of surf beat. That is, the bundle of energy in a package of several higher waves, which cause a short-term rise in water level on a beach, may be passed along. It would be difficult to draw meaningful orthogonals for this long-period disturbance (the period is about half that of surf beat) to determine where the next bounce would come. In certain areas they may be responsible for special conditions of erosion, sediment deposition, and the existence of rhythmic features called *cusps*.

STORM SURGES

During a violent storm there may be a substantial rise in the sea level along a coast that is known as a storm tide or storm surge when combined with the astronomical tide. When this happens, the wind-raised storm waves are superimposed on this surge, and sometimes the land is invaded with disastrous effect. This rise in water level is often the combined result of an atmospheric (barometric) low-pressure area surrounded by high-pressure areas offshore; that is, the differences in air pressure cause a hump in the sea surface under the low-pressure area. As the storm moves toward land, this hump of water invades coastal areas. In addition, the strong winds create large waves and drive them across shoaling water, piling them on the shore. These steep, breaking waves are driven so hard, one on another, that they create a general landward-flowing surface current (mass transport plus translation) that moves faster than the surplus water can return seaward along the bottom. While all this is going on at sea, heavy rains usually pummel the land, raising the levels of the streams and rivers. The rainwater, driven by gravity, attempts to flow into the already-rising sea. The result of these three factors is a flooded coast, and the battering waves atop the general flood cause the most serious destruction, for shoreline structures are now in the surf zone.

Some case histories of the great destructive effect of such surges serve to remind us of how quietly the ocean usually lies in its basin and how damaging a small change in sea level can be—regardless of origin. The oldest known inhabited site to have been inundated by a storm surge was at the archaeological site of Lothal, India, about 4,000 years ago. A more recent event was the Galveston, Texas, "flood" of 1900. On that occasion a hurricane with winds of 120 mph (190 km/h) raised the water level along that shore of the Gulf of Mexico 15 feet (5 m) above the usual 2-foot (60-cm) tidal range. The storm waves, probably another 25 feet (7 m) high, rode in atop the storm tide and demolished the city; some 5,000 people drowned. Figure 25 shows another example of sea level rise during a storm surge from Puerto Rico.

On February 1, 1953, a strong gale swept down the North Sea and piled its waters against the Dutch coast. The storm surge rose 10 feet (3 m) above the highest high-tide level; the waves topped the dikes. The overflowing water eroded gullies in the unsheathed inner side of the dikes until they were breached in sixty-seven places. In a short time, channels up to 100 feet (30 m) deep and 1,500 feet (450 m) wide were cut in the dikes. As the sea defenses collapsed, the North Sea poured in,

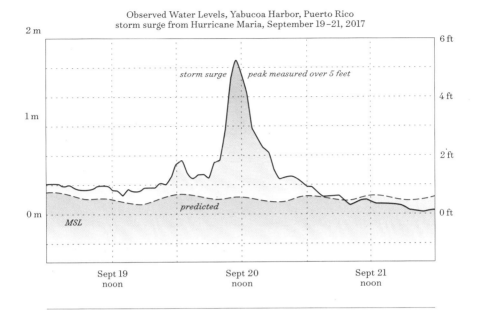

Observed Water Levels, Yabucoa Harbor, Puerto Rico
storm surge from Hurricane Maria, September 19–21, 2017

2 m

6 ft

storm surge   *peak measured over 5 feet*

4 ft

1 m

2 ft

predicted

0 m

0 ft

MSL

Sept 19
noon

Sept 20
noon

Sept 21
noon

FIGURE 25: Hurricane Maria's storm surge was created by intense winds, waves, torrential rains, and runoff from the land. It peaked near 6 feet (2 m) on a coast accustomed to a 1-foot (0.3-m) tide. It is the most costly Caribbean hurricane on record.

and a steep-fronted wave advanced across the low country. The result was that 800,000 acres (324,000 hectares) were flooded, 1,783 people drowned, and the damage exceeded $250 million USD. But in a few years the Dutch had repaired the dikes, reclaimed their land, and were extending the dike system to include new lands. This super-gale was later determined by Dutch engineers to have been a "400-year storm." That is, the chance that all the unfavorable circumstances (high winds of long duration in that direction) would occur at one time was such that there was a likelihood of similar occurrence only once every 400 years. Today the Dutch are the masters of shore protection. They have built immense structures to withstand storm surges and protect their citizens. These structures and legal framework changes will be discussed in Chapter Twelve.

In Bangladesh and India, the low flat shores along the Bay of Bengal are densely populated and have at various times been invaded by great storm surges, which have taken a terrible toll of life. In 1876,

Storm surges and astronomical tides in the Adriatic Sea contribute to the *acqua alta* (high water) in Venice, Italy. *Marco Secchi/Corbis via Getty Images*

the rising water caused more than 200,000 deaths. On October 31, 1960, at Chittagong, East Pakistan (today Bangladesh), near the mouth of the Ganges, a cyclone and storm surge caused much damage. Three small islands were completely inundated and thousands of lives and houses were lost. Then, in November 1970, tropical cyclone Bhola, perhaps the deadliest storm of the twentieth century, struck Bangladesh. As many as 500,000 people perished in the storm—no one knows for sure—and entire villages and towns disappeared in the waters that rose an estimated 33 feet (10 m) in some areas. The effects of Bhola and its subsequent disaster management provide a study of how the forces of nature changed the geopolitics of the region. Our changing climate will play an increasing role in local and global politics as our low-lying cities attempt to combat storm surges superimposed on a rising sea level. Construction began on a $6 billion USD storm surge barrier for Venice, Italy, in 2003. The project, called Mose, is scheduled to be completed in 2022. Stormwater surged into Venice during November 2019, reaching the second highest water level since official records began in 1923. Venice is sinking while sea levels continue to rise.

The New England coast of the United States has had many serious storm surges. During Hurricane Carol in August 1954, the water level in Providence, Rhode Island, rose 16 feet (4.9 m) above normal, flooding the downtown business area 8 feet (2.4 m) deep and causing losses of many millions of dollars. Providence city officials took action in the early 1960s and built a protective tidal barrier rising 22 feet (6.5 m) above average sea level across the upper end of Narragansett Bay. Although the structure cost many millions to build, storm surge damage has been greatly reduced over the years.

The Wave Changing Us All

Sea-level rise is a global problem—the forces of nature do not stop at territorial boundaries. Storms of great intensity have increased in the twenty-first century. Hurricane Katrina affected the entire Gulf of Mexico in August 2005. Katrina was a Category 5 storm with winds reaching over 170 mph (270 km/h) with a storm surge of more than 32 feet (10 m). The low-lying areas of Louisiana and Mississippi were inundated as far as 6 miles (10 km) inland; over thirty offshore oil platforms were destroyed or damaged; and nine coastal refineries were closed, resulting in many oil spills. The devastation caused an unprecedented spike in oil prices and prompted then-President George W. Bush to release over 11 million barrels (1.7 million m³) from the Strategic Petroleum Reserve. Hurricane Katrina caused more than $100 billion in damage—it was perhaps the most costly (in terms of financial loss) natural disaster in the history of the United States at the time. In New Orleans, where the surge protection system and levees failed, 80 percent of the city and surrounding areas were flooded. Many of these losses, including some of the 1,000 or more lives, could have been spared if the public and governments had been better prepared. Katrina also generated a tempest of political problems for any climate change denial politicians attempting to appease the rising flood of climate activists.

Hurricane Sandy struck the eastern coast of the United States on October 29, 2012. Although the loss of life was relatively low (fewer than 200 people), the damage to property has been estimated at over $50 billion, a cost borne by individuals, insurance companies, and governments. Sandy destroyed or damaged more than a half million homes and resulted in power outages for more than 8 million people. As a result of this hurricane, many insurance companies restructured their actuary tables (their basis for charging customers) and future loss projections. Also unprecedented were the rain and flooding near Baton Rouge, Louisiana, in 2016. These two flooding events indicate a new normal.

The Atlantic hurricane season of 2017 was of exceptional intensity with ten hurricanes of Category 3 (winds over 110 mph (177 km/h)). Hurricanes Harvey, Irma, and Maria collectively caused most of the more than $300 billion USD in damage. In late August 2017, a tropical storm strengthened into the Category 4 Hurricane Harvey and made landfall in southern Texas. The ensuing rains were greater than any in the recorded history for the area. The heavy rains were statistically a 1-in-10,000-year rain event.

Hurricane Maria devastated many Caribbean islands with intense rains and storm surges in September 2017. Maria was the worst natural disaster in Puerto Rico's history, costing an estimated $80 billion USD, and resulted in approximately 3,000 fatalities. Maria destroyed Puerto Rico's electrical grid and 80 percent of its agricultural crops. At Yabucoa Harbor, the storm surge reached 6 feet (2 m) above normal around 8 am on September 20, 2017. The combination of rains, storm surges, and winds of up to 175 mph (280 km/h) completely halted commerce. Just two days prior on September 18, 2017, Maria made landfall in Dominica. Already suffering from heavy rains and landslides, the Caribbean island churned with destruction—even the prime minister needed to be evacuated from the official residence. The aftermath of Hurricane Maria is still echoing with political waves, some still unresolved, storm-surging into the highest offices of the US government.

In September 2019, the Bahamas were struck by Hurricane Dorian, its most intense and damaging hurricane on record. Dorian packed 185-mph (300-km/h) winds, heavy rains, 910 millibar low pressure, and storm surges, and initiated a state of emergency up the east coast of North America. The hurricane cost billions of dollars in damage. Clearly the nature of hurricanes has changed.

Hurricanes of the Atlantic are not alone in their devastation; in the northwestern Pacific Basin a mature tropical cyclone is called a typhoon. There have been many notable examples of typhoons in recorded history. One example is the Asian Super Typhoon Haiyan (known as "Super Typhoon Yolanda" in the Philippines), which caused over 6,000 deaths and $3 billion USD in damage. In November 2013, Haiyan's storm waters surged to almost 23 feet (7 m), infiltrating and contaminating crucial freshwater aquifers. This impact—the storm surge-related contamination of aquifers by seawater—is poorly reported. Storm surges and sea-level rise that are due to climate change both threaten coastal aquifers. A small increase in sea level has already rendered many freshwater supplies unusable for agriculture and human consumption.

These examples are intended to provide an introduction to how our environment is changing. Intense atmospheric winds have created a new palate of waves upon a rising sea. The world's low-lying coastal regions and beaches have been encroached on by cities, driven by human population growth. Approximately 10 percent of the world's population lives below 33 feet (10 m) elevation. In Vietnam it is more than 60 percent of its population. Neither the beaches nor the low-lying cities will remain unchanged.

There is no question that we are experiencing a period of rapid environmental change, which has been observed by residents around the globe and confirmed by the data collected by scientists. Today's geological and geochemical methods have identified periods of great change many thousands of years ago, many of which foreshadowed catastrophic events. Humanity has enjoyed about 4,000 years of relatively stable climate conditions as indicated by that period's stable sea level. The same climate stability that allowed our civilizations to develop is now uncertain.

Climate change is a wave affecting the water we drink, how we grow food, where we build our houses, our tax revenues, the design and strength of our ships, which products we manufacture, how we distribute goods, and how we provide services. The derivative industries of property insurance, risk mitigation, and disaster relief must develop new metrics and methods of survival. In the twenty-first century, climate and geopolitics will evolve in symbiosis, or risk deteriorating into chaos.

Super Typhoon Haiyan's storm surges reached 23 feet (7 m) tall. Driven by peak winds of 200 mph (300 km/h), Haiyan blended sea, land, and man-made structures into the devastation shown here. Guiuan, Philippines. *Bryan Denton/The New York Times via Redux Pictures*

5 Winds and Waves of Climate Change

Some things change unnoticed; others transform endlessly, forever. The Earth has been experiencing measurable changes in atmospheric weather patterns, ocean temperatures, intensity of storms, and precipitation since records began. Although complex in origin, the results of these changes are easily observed by the nonscientist. Recently, in both 2017 and 2018, we saw an increase in the intensity of Atlantic tropical storms and flooding. Indeed, the 2017 total accumulated cyclone energy (ACE is a measurement related to hurricane strength and duration) was deemed "hyperactive" and in 2018 "above normal" by the National Oceanic and Atmospheric Administration.

How does this effect the beach? The results of higher ACEs can be compared to the higher settings on a kitchen blender. The kitchen blender's settings: low-level "pulse," medium-level "stir," and highest-level "whip" have different results. A low-level pulse of energy from the atmosphere into the ocean is like a short storm, it causes disruptions, yet things return to rest soon after. The next, medium level is

PREVIOUS SPREAD: Typhoon Hagibis came ashore at Mihama, Mie Prefecture, Japan, in October, 2019, with abundant energy. *Franck Robichon/EPA-EFE* like the stirring energy of a tropical cyclone causing changes that do not easily return to rest. The highest levels of ACE, the extreme Category 5 cyclones, whip energy into the ocean, resulting in permanent changes to bars, berms, and beaches. A new normal has been established—neither the coastline nor the contents of the blender will ever be the same.

Sadly, it has become an irreversible process. Any increase in the speed or the duration of the wind will cause larger waves to form, and the increased energy stored in the waves will be poured devastatingly upon the land. Far above the sea and the land, the meanders of the jet stream (high-altitude, fast winds) now embrace newfound energies from a warming planet. We experience these larger meanders in our weather in that some areas are suddenly warmer, then abruptly colder, responding to the jet stream's wanderings, changing the patterns of our lives. These longer-term dynamics are complex and far reaching. In the polar regions, wind-blown dust settling on glaciers has changed the amount of light reflected by the Earth (known as the *albedo*) as has algae growing on the underside of ice. Dust plays another important role; it provides iron for the world's oceans, stimulating plankton growth and carbon dioxide uptake. These are glimpses of the onrush of winds and waves of climate change.

SEA LEVEL

No portion of the Earth can avoid the effects of climate change. The Pacific coral atoll island nation of Kiribati, once believed to be far from the complexities of twenty-first century urban life, is being threatened by rising sea level and changing weather patterns. Kiribati's seawater dynamics and freshwater resources have been in a delicate balance since first being settled by *Homo sapiens* about 1,000 years ago. The slow but continual formation of a healthy coral reef depends on many things, including stable ocean water temperatures, prevailing wind and wave directions, and its spur-and-groove alignment. These finger-like forms extend toward the waves, directing the flow of seawater over the reef and then into the lagoon. Stable weather patterns have provided protection from waves and have maintained Kiribati's freshwater dynamics. The combined effects of rising sea level, saltwater intrusion, coral bleaching, and changing weather patterns have managed to upset Kiribati's dynamics. Up to 80 percent of the reef surrounding Kiribati has not recovered from the 2015–2016 coral bleaching event, and the wave climate surrounding Kiribati has intensified, offering the land no choice but to surrender through erosion and submersion.

Kiribati survived the bloody Battle of Tarawa in 1943 between Japan and the United States, and nuclear weapons testing during the 1950s and 1960s, but it may be the first nation not to survive the twenty-first century due to climate change. Kiribati is not alone. It shares its fate with Tuvalu, the Marshall Islands, the Maldives, and innumerable low-lying coastal communities in the Netherlands, Japan, Vietnam, Thailand, and Bangladesh. The Carteret Atoll, off the coast of Papua New Guinea, has had population relocations due to changing environmental conditions for more than a decade.

In the United States, rising sea levels threaten many major cities, including Boston and New York. Much of the Outer Banks, the Intracoastal Waterway, barrier islands, Gulf Coast (Florida to Texas), and the Pacific coastline from Canada to Mexico must confront the rising seas. In Pacifica, California, cliff erosion has already led to the destruction of many houses. Further south in the state, the city councils in Oceanside, Carlsbad, Del Mar, and San Diego have addressed the integrity of coastal seawalls, roads, and bridges, all threatened by the wave of climate change. Some municipalities have chosen not to battle the sea—and opted for a "planned retreat" from rising waters. Others, like in North Carolina where they passed HB 819 in 2012, have strangely banned policies from being based on climate change predictions. Unfortunately, six years later, in September 2018, Category 4 Hurricane Florence produced large storm surges and inundated much of the Carolina coastline and the offshore islands. North Carolina may have proved it can ignore the findings of climate scientists, but it cannot escape the impacts of climate change. Unfortunately for the US citizens living in these coastal areas, the US federal government has yet to provide a national policy on coastal climate change. Nevertheless, sea-level changes will mobilize some labor pools, because workers are more mobile than factories.

The intricate, finger-like coral ridges and channels (spurs and grooves) seen in the foreground are aligned with the local predominant wave direction. Tubbataha Reef Natural Park, Philippines. *Tommy Schultz*

In many other countries, including the United Kingdom, climate change is being maturely addressed by local councils and environmental agencies. Even the Isle of Wight (population 140,000) has its own Shoreline Management Plan for climate change and the implications of sea-level rise. These devastating changes are not limited to the mid-latitudes and tropics. The Arctic, Bering, Chukchi, Beaufort, and East Siberian Seas (between the United States, Canada, and Russia) are all experiencing change. Because of warmer water temperatures, sea ice is decreasing and exposing more open water. The decrease in sea ice now allows the wind to blow over greater distances, forming bigger waves that attack the previously ice-protected shores. This is especially the case for the west coast of Alaska near Nome. Where the ice previously

Immense chunks of permafrost, frozen for millennia, thaw and collapse
into the Arctic Ocean. The exposed coastline is further undermined
and eroded by waves across the ice-free water. Alaska's Arctic Coast.
Christopher Arp/USGS

protected the shores in Nome and across the Arctic, the wave climate
is changing, erosion is on the rise, permafrost is thawing, and coastal
flooding is more frequent.

Walter Munk, the researcher who assisted in the wave forecasting for D-Day in 1944, has provided the scientific world and general
public with insight and wisdom on many topics. Recently, in October
2017, Munk's 100th birthday was celebrated in San Diego, California. To commemorate his centennial, local community leaders and
elected officials unveiled Walter Munk Way near the shoreline in the
coastal community of La Jolla. Walter's words were wise yet forebod-
ing as he said, "Unless we have an organized and concerted effort to
reverse the human-caused results of climate change, unfortunately

'Walter Munk Way,' unlike me, will not see its 100th birthday—it will be underwater."

There is an interesting legal aspect to rising sea levels. Many legal jurisdictions define "public waterways" in relation to the local water levels, mean sea level, and tides. As sea level rises, many privately owned land parcels have fallen into the legal definition of a public waterway. The repossession of private land by the government is not hypothetical—several southern US state governments have already successfully enforced claims to private land. Private landowners in lowlands will continue to lose their property and mineral rights, not just to climate change but also to government entities. Indeed, human-caused climate change will reposition our beaches, coastlines, aquifers, and waterways at rates unseen for thousands of years and never before experienced by human civilizations.

Harnessing the Power of Waves

The thunder of a large wave breaking is a release of a huge amount of stored energy. A reasonably steady flow of this energy exists in the trade wind seas just north and south of the equator. There, during most of the year, waves roll steadily across the lower latitudes at an average height of about 9 feet (3 m) and a period of about eight seconds. In such trade wind seas, the power expended (or, the amount of energy expended per unit time) by the waves in raising and lowering a passing ship every eight seconds is greater than the power of the ship's engines.

All sorts of inventions have been created to harness the waves (see figure 26, page 126). A wave-powered pump was invented by John Isaacs (1970s), a prototype that worked very well, even in low waves. It used water to compress air; and a turbine atop the buoy converted the pumped water into electricity. What was envisioned in the 1970s has become a commercial reality. In 2005, the construction phase began on a commercial wave energy farm 3 miles (5 km) off the Portuguese coast. The Aguçadoura Wave Farm was projected to cost more than $10 million and was designed to produce 2.25 megawatts of power, enough for hundreds of homes. The project was deployed and functioned during 2008 but was "retired" that same year due to structural problems.

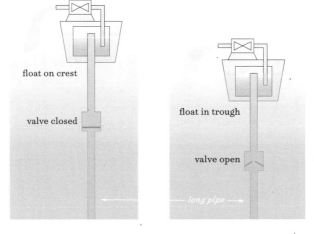

FIGURE 26: The Isaacs wave-energy pump uses wave motion to power a turbine. The descending buoy forces water up a long pipe through a one-way valve. Pressure in the air chamber keeps the water flowing steadily through a turbine.

Today, other systems abound, ranging from the small-scale Ocean Motion Technologies, founded by a Scripps Institution of Oceanography student, to several multimillion-dollar international investments for grid-tied solutions.

The US Bureau of Ocean Energy Management (BOEM) and others provide lists of projects. BOEM estimates the amount of "recoverable energy" within US jurisdiction is more than 2 terawatt hours per year; almost one-third of the current US energy demand. A small number of wave power installations are in operation around the world.

INTERACTIONS OF ATMOSPHERE, OCEANS, AND HUMANS

Immediate and long-term actions are needed to resolve problems of global greenhouse gas imbalances due to the burning of fossil fuels. An increase of atmospheric carbon dioxide is directly related to a warming Earth. The oceans are responding by absorbing about 25 to 30 percent of our atmospheric oversupply of carbon dioxide and most (about 90 percent) of the excess heat. These excesses have accompanied a sea-level rise of roughly 0.5 feet (15 cm) in the past 100 years.

Unfortunately, the rate at which humans are producing carbon dioxide is still increasing, as are the rates of ocean and atmospheric warming. In 2015, the global atmospheric carbon dioxide level surpassed 400 parts per million (ppm)—30 percent higher than it has ever been during the past *several million* years. Because of the added heat, we are experiencing new levels of turbulent storms, winds, and waves. Wind waves, sea state, and whitecaps are important drivers for the exchange of atmospheric gases into the ocean. Warmer ocean temperatures supply us with sea-level rise—a love letter in a bottle from the melting of glaciers and polar ice caps. Today our offshore activities and structures are only partially regulated by the United Nations Convention on the Law of the Sea (UNCLOS, initiated by Maltese diplomat Arvid Pardo). Enforcement of the UNCLOS laws is driven by individual national interests. There is hope for an enforceable legal framework, as new international patent law includes the category Y02 defined as "Technologies or Applications for Mitigation or Adaptation Against Climate Change." Intellectual property violations might be another basis to bring legal action against those who cause climate change. How do we combat our own actions on a global scale, with voluntary restraint or imposed regulations? It may be the most existential question ever asked.

The Europeans have been the leaders in harvesting renewable energy and implementing energy-conservation methods. Recently, Britain, Denmark, Holland, and Sweden have all put ocean-based, commercially competitive, renewable energy systems into operation (the Chinese have invested more in photovoltaic research, development, and production than any other country and more than all of Europe combined). Maddeningly, the United States lags far behind the rest of the developed world in energy efficiency, whereas the US per capita use of energy is among the highest.

We must press for decreasing the per capita energy use, do more research on non-fossil-based sources of energy, and increase the efficiency of current technologies, while at the same time increasing conservation efforts. Limiting the number of children per family is not a popular topic, but there is a global wave of human population growth. For some, limiting population may be a religious taboo or GDP heresy. However, if we continue to let our population grow as it has over the past 150 years, no amount of green or other energy will spare us from environmental collapse. Instead of buying a new car every two years, perhaps we can change the ethos and each produce one child per generation. If we have any, our grandchildren will appreciate a less politically turbulent world, fewer energy wars, a cleaner environment, stabilized greenhouse gases, and abundant renewable resources.

The elusive winds and waves of climate change have recast coastal dynamics; now clearly visible in scientific data covering periods of decades, centuries, and millennia. There are other types of waves which cause the sea level to change for periods of minutes to hours. In the next chapter we will look at two types of waves which have well-understood origins and predictable periods, these are tides and seiches.

Kansai International Airport was inundated and shut down by Typhoon Jebi in 2018. Thousands of passengers were stranded at the airport (built on an artificial island) in the middle of Osaka Bay. Jebi was the most powerful typhoon to strike Japan in 25 years, packing winds of 175 mph (280 km/h). *The Asahi Shimbun via Getty Images*

6   Tides and Seiches

The study of the tides is a large and complicated subject, most of which is beyond the scope of this book. The tides, however, are an important form of long-period wave, and it would be illogical to ignore them entirely. Besides, they play an important part in beach and coastal processes because they constantly change the depth of water in which waves approach the coast and the level at which waves strike the beach. Therefore, in this chapter we shall touch lightly on the main points and encourage the especially interested reader to dig into the references (at the end of the book) for more detailed information and a fuller explanation.

THE TIDES

On all seacoasts, there is a rhythmic rise and fall of the water called the *tide*, and associated with this vertical movement of the water surface are horizontal motions of the water known as tidal currents. Together they are known as tides.

PREVIOUS SPREAD: Low tide at Tobermory on the Isle of Mull, Inner Hebrides, Scotland. The rise and fall of waters have created empires and lost wars. *Feifei Cui-Paoluzzo/Getty Images*

Tides are the longest waves oceanographers commonly deal with, having a mean period of 43,000 seconds (twelve hours and twenty-five minutes) and a wave length of half the circumference of the Earth. The crest and trough of the wave are known respectively as *high tide* and *low tide*. The wave height is called the *range of tide*, but because it is measured in places where it is influenced by the shape of the shore, it varies greatly from place to place.

The gravitational attraction of the Moon and Sun on the Earth and its waters cause tides. Long before a word for gravity existed, the ancients realized that there was some connection between the Moon and the motion of the water. Western civilizations developed on the shores of the Mediterranean, an essentially tideless sea where the winds and the barometric pressure frequently played the greater roles. Not until a number of explorers had ventured beyond the Pillars of Hercules into the Atlantic and observed tides, where the range is large, was the relationship between the phases of the Moon and the height of the tide established. Then some 1,500 years passed before Johannes Kepler wrote of "some kind of magnetic attraction between the Moon and the Earth's waters"—and Galileo scoffed.

It remained for Isaac Newton to discover the law of gravity, which holds that the gravitational attraction between two objects is directly proportional to their masses and inversely proportional to the square of the distance between them. From this relationship it can be shown that the gravitational attraction of the Sun for the Earth is about 150 times that of the Moon. The tremendous mass of the Sun more than makes up for its much greater distance. But the Moon is the primary cause of tides. Why?

The answer is that the *difference* in attraction for water particles at various places on the Earth is far more important than total attraction. That is, because of the Moon's nearness (on average only 239,000 miles (385,000 km)—more about lunar distances later in this chapter) there is a big difference in the lunar gravitational attraction from one side of the Earth to the other.

The water on the side of the Earth nearest the Moon is some 8,000 miles (13,000 km) closer to the Moon than on the far side. The Sun, however, is 93 million miles (150 million km) away, and a few

HIGH TIDE The high tide is the result of a maximum in astronomical forces. Bedruthan Steps, Cornwall, United Kingdom. *Michael Marten*

thousand miles one way or the other make comparatively little difference. Thus, the Sun's gravitational force, although far larger, does not change very much from one side of the Earth to the other. So the Moon is more important in producing tides. For the sake of simplicity much of the following discussion speaks only of the effect of the Moon, but the Sun's effect is similar.

The result of these differences in gravitational attraction is that two bulges of water are formed on the Earth's surface: one toward the Moon, and the other, as we shall see, away from it. The Earth rotates on its axis once a day, and it is not difficult to imagine that it turns constantly with a fluid envelope of ocean whose watery bulges are supported by the Moon. This concept considers the tide wave to be standing still while the ocean basin turns beneath it. Thus, most points on Earth experience two high tides and two low tides each day.

It is easy to see why the gravitational attraction of the Moon should raise a bulge of water on the side of the Earth toward the Moon, but it is not quite so easy to understand why there should be a similar bulge on the opposite side, away from the Moon. Let us try to clarify that point.

LOW TIDE Low tide is the result of a minimum in astronomical forces. The water receded 25 feet (8 m) from high tide. Bedruthan Steps, Cornwall, United Kingdom. *Michael Marten*

The Earth and the Moon revolve *around each other*, around a common center of mass, roughly once a month (see figure 27, page 136). Surprisingly, the center of the monthly Earth-Moon revolution (their *barycenter*) is not at the center of the Earth, but is always offset toward the Moon by roughly 2,900 miles (4,600 km).

This whirling Earth-Moon system creates the outwardly directed centrifugal forces, which get entwined in gravitational tidal forces. The side of the Earth nearest to the Moon experiences larger lunar gravitational forces while the far side of the Earth is experiencing lesser lunar gravitational forces.

Each bulge (near side and far side) is a result of the local combined gravitational and centrifugal forces. At a fixed point on the surface of the Earth (or ocean), both these forces are constantly changing: gravitational because of the varying distance to the Moon and centrifugal because of the varying distance from the barycenter. These changing forces cause the constant movement and redistribution of water known as the rise and fall of tides. The length of each of the arrows shown in figure 27 (page 136) corresponds to the magnitude and direction of the

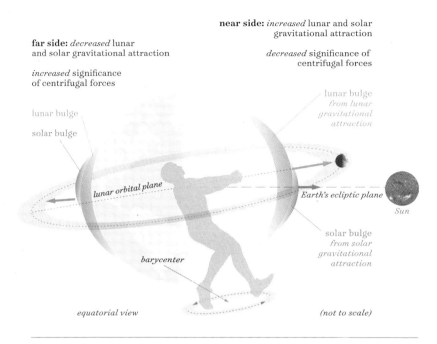

far side: *decreased* lunar and solar gravitational attraction

increased significance of centrifugal forces

lunar bulge

solar bulge

lunar orbital plane

barycenter

equatorial view

near side: *increased* lunar and solar gravitational attraction

decreased significance of centrifugal forces

lunar bulge *from lunar gravitational attraction*

Earth's ecliptic plane

Sun

solar bulge *from solar gravitational attraction*

(not to scale)

FIGURE 27: Tides are caused by slight local changes in the gravitational attraction between orbiting bodies. Attraction is the greatest at the point where the Moon (or Sun) is the closest. However, on the far side of the Earth, the gravitational attraction is reduced, and the less-encumbered, whirling centrifugal forces bulge the water outward.

combined forces which produce the tidal bulges on the near and the far sides of the Earth.

To help us understand the tidal bulge related to centrifugal force, let us use the rotating hammer-thrower as an example, as shown in figure 27. The center of rotation is at the feet of the hammer-thrower. The arms are pulled outward in one direction (toward the Moon) as the hammer-thrower's body and head lean back in the opposite direction. The thrower's head is pulled backward by a centrifugal force, in the opposite direction to the hammer weight. The loads of the Moon and Sun's gravitational forces are diminished on the far side of the Earth (farthest from the Moon). There the unburdened centrifugal forces emerge as a far-side tidal bulge.

Remember, the Earth and the Moon revolve around their center of mass once a month. A lunar month (also known as a synodic month),

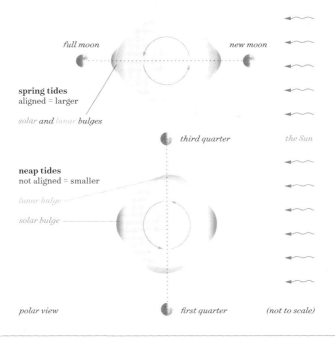

full moon

new moon

spring tides
aligned = larger

solar **and** *lunar* *bulges*

third quarter

the Sun

neap tides
not aligned = smaller

lunar bulge

solar bulge

polar view

first quarter

(not to scale)

FIGURE 28: Lunar and solar gravitational forces work together when aligned, and compete when not aligned. Thus, periods of larger tides (aligned for spring tides) and periods with smaller tides (neap tides) occur over a lunar month (29.53 days).

the period of time that passes between two successive full Moons or new Moons, has a long-term average of 29.53 days.

The Moon orbits in the same direction as the Earth spins about its own axis. But they move at different rates, which causes the Moon to be retarded toward the east, rising later by about 50 minutes every day. This motion of the Moon requires any point on Earth to go slightly farther than one revolution to come beneath the Moon again; thus, the *tidal day* is longer than twenty-four hours. It is twenty-four hours and fifty minutes long.

One further complexity is that the lunar tidal bulge is rarely directly beneath the Moon, as shown in figure 28 above. The lunar bulge follows the Moon. This is the result of the friction of the Earth as it rotates beneath the water. The rough-bottomed ocean basins tend to drag the bulges along; the gravitational effect of the Moon tends to hold its bulge beneath it. Hence, gravity and friction attempt an equilibrium. In consequence, high tide arrives after the Moon (called phase lag) has passed a point on Earth (meridian passing). However the solar bulge can influence the arrival of the lunar high tide (referred to as priming or lagging of the tide).

The Sun tides, though much smaller, are important because of the way they increase and reduce the lunar tides. The two most important situations are when the Earth, Sun, and Moon are aligned (in phase) and when the three make a right angle (out of phase). In the in-phase case the solar bulge rides on top of the lunar bulge to make *spring tides*. During spring tides, which have nothing to do with the spring season but occur about every two weeks, the water level rises higher and falls lower than usual. This large range of tide lasts two or three days; then the two bulges get progressively further out of phase until a week later, when the high and low tides are about 20 percent less than average. These are *neap tides*; in effect, the Sun's gravitational force reduces the Moon's bulges. Between these two extremes the solar bulge contributes in a way that warps the shape of the main bulge, and the high and low tides come a little earlier or later, slightly varying the length of the tidal day. Armed with this information, we are much better equipped than in ancient times to look at the Moon and forecast the height of the tide. At new Moon and full Moon there are spring tides; neaps come when the Moon is in the first or last quarter (see figure 28, page 137).

Another important variation in the height of the tide is the result of the Moon's elliptical orbit about the Earth (see figure 27, page 136). The Moon's orbital distance from the Earth varies roughly between 225,000 and 250,000 miles (362,000 and 405,000 km). At *perigee* (the nearest point in its orbit), the Moon is 15,000 miles (24,000 km) closer; at *apogee* (the furthest point) it is that much farther away. This change in distance (and therefore in the attractive force) causes tides that are, respectively, 20 percent higher and lower than average. Moon-Earth perigee is reached once per orbit (once a month) and only rarely does this coincide with the in-phase alignment of Sun, Earth, and Moon (see figures 28 and 29). But at least twice a year both effects exist at the same time—that is, a full Moon (or a new Moon) exists when the Earth is closest to the Sun (*perihelion*). Then the Sun's closeness adds to the

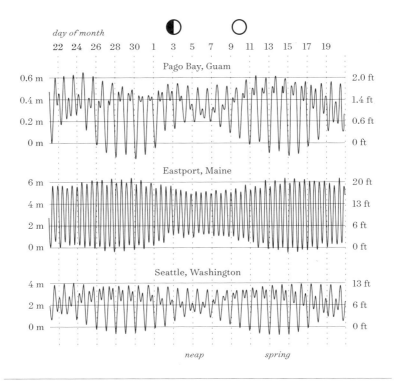

FIGURE 29: Tidal records (from three locations) each show large inequities in water levels during a lunar month. During the Moon's first quarter, smaller (neap) tides arise because the Moon is not aligned with the Sun. A week later, during the few days near the full Moon, larger (spring) tides occur (see figure. 28).

lunar spring tides to produce the highest tides of the year. A lunar perigee can provide us with the so-called "supermoon"—because the Moon is "super close." The greatest tides occur when the Moon is the closest to the Earth while the Earth is closest to the Sun (early January). In some countries, these tides are called *king tides*.

If you take the time, you might notice that the Moon rarely rises at the same location. Why is this so? The Earth's path around the Sun is called its *ecliptic* and the Earth's axis is tilted (or, slanted) by 23.4 degrees relative to the ecliptic. Astronomers call this tilt the *obliquity of the ecliptic*. Likewise, the Moon has an inclined orbit relative to the Earth's ecliptic of about 5 degrees (see figure 27, page 136). Because of all these tilts and relative motions, you can observe moonrise further north or further south as it rises into view above the eastern horizon. Some tidal influences take more than eighteen years to repeat. All these positional complexities, as you might have guessed, each affect the tide. We will leave the finer facets of tidal forces to the tidal experts; you will find a few references in the Further Reading section of the book.

Having considered the main forces that produce tides, we now can think about how these curious waves behave. Many shores, including the US Atlantic coast, experience two tides a day of about equal heights; these are called *semidiurnal* (semi-daily). A few places in the world have only one high and one low a day. And most of the Pacific and Indian Oceans have mixed tides; that is, the heights of high and low waters are unequal, as in Seattle, Washington, or Pago Bay, Guam (see figure 29, page 139).

The cause of this changing inequality occurs because when the Moon is opposite the equator, the highs and lows are about equal. But when the Moon is "in the tropics"—that is, above the Tropics of Cancer or Capricorn—different thicknesses of bulge move past points away from the equator. The Earth's tilt complicates the ocean's changing tidal inequality.

One of the most important influences on the height and character of the tide is the shape of the basin where it is observed. Measurements of the height of the tide in the deep ocean have been made. There the range is typically small, about 1 foot (30 cm), as it is on mid-ocean islands, such as Hawai'i. But as the solid Earth turns beneath the tidal bulge, the shallow continental shelf acts as though it were a wedge driven under the wave front. The result is that the deep-water tidal range is much exaggerated at the shore. Estuaries with wide, funnel-like openings to the ocean tend to amplify the tide range further. The width of the tide wave that enters the opening is restricted as the channel narrows; this constriction concentrates the wave energy and increases its height. Of course, if the estuary is very long, the frictional effects of sides and bottoms gradually reduce the height of the tide wave until it vanishes.

The importance of coastal configuration is illustrated by the difference in tidal height between Nantucket Island, about 1 foot (0.3 m), and the Bay of Fundy with more than 50 feet (15 m), which are only a few hundred miles apart. The opposite ends of the Panama Canal are only about 50 airline miles (80 km) apart, but there is a great difference in the tides at the two terminals. At Colón, on the Caribbean side, the tide is generally diurnal and the range is about 1 foot (0.3 m); at Balboa, on the Pacific side, the tides are semidiurnal with an average height of 14 feet (4.2 m), and the locks are built to withstand spring tides of as much as 21 feet (6.5 m).

Although the tide doubtless advances across the ocean like a sine wave, there are only a few places on Earth where this pattern can be observed directly. One such location is Chesapeake Bay, which has its opening to the east then extends to the north. There the troughs and crests (the low and high tides) slowly move up the long shallow bay as a series of "progressive waves." Usually there are two high-tide zones within its 150-mile (240-km) length at the same time, with 50 miles (80 km) of low-tide water in between.

With the aid of satellites, tides have been observed in the Arctic and Antarctic influencing the massive ice shelves. The polar tidal forces combine with wind-driven waves to break up ice shelves and move the sea ice. The Antarctic Ocean is a continuous belt of water extending completely around the Earth, 600 miles (1,000 km) wide at its narrowest point. The tidal bulges must act as a forced wave into which energy is constantly being added by the tide-producing forces. As each high tide passes the openings into the Indian, South Atlantic, and South Pacific Oceans, waves are initiated that travel freely northward and modify the local tides as they go.

Accompanying the rise and fall of the tide are substantial horizontal motions of water known as tidal currents. Like the vertical changes, they have little significance in the open sea, but in harbors and narrow estuaries they are of considerable importance. On a rising tide the currents are said to be *flooding*; on a falling tide, they *ebb*. The direction of flow is the set of the current.

When there is no flow, the current is slack; thus the time of slack water is usually within an hour of high or low water. The maximum current velocity comes at about the same time as the maximum rate of change in the height of the water. Because these relationships depend largely on the local conditions, no general rule applies.

Tidal currents of 15 knots in Seymour Narrows, British Columbia, Canada, and 4 knots in the Golden Gate to San Francisco Bay are normal. Such currents have little influence on open beaches, but they may

The Seymour Narrows in British Columbia, Canada, experiences tidal currents reaching 15 knots, driven by the 9-foot (3-m) tides. *Phillip Colla*

have considerable effect on sand movements in and near harbor mouths. For example, the combined effect of these currents and of ocean wave forces often causes a sandy bar (or, harbor bar) to form just outside the harbor entrance. This barrier will cause large waves to break and endanger small craft entering or leaving the harbor, thus explaining the many allusions in literature to sailors' fears of a breaking or *moaning bar*. Tidal or other currents will cause waves to break or shorten their wave length. Such changes in the texture, color, temperature, and salinity of the sea surface permit the Gulf Stream and other great currents to be recognized from aircraft and spacecraft.

Local sea level is the height that the sea surface would assume at one location if it were undisturbed by waves, tides, or winds. But because these disturbances do exist, the technique of averaging all possible sea levels has been adopted. The result is *mean sea level*, a convenient datum plane from which heights of tide or depths of water on a chart are measured. Charts and tide tables for the US Pacific coast refer to another datum: *mean of lower low waters*—the average height of the lowest of two low tides a day. Mean sea level for the entire Earth is similar but much more complicated. *Global sea level* takes into account the variations in the Earth's gravity field and is referred to as the Earth's "geoid." Mean sea level has been rising because of a warming climate. Seawater expands as it gets warmer. Additionally, during our current period of ice sheet decay, meltwater is flowing into the ocean basins, transferring mass out of the polar regions and raising the global sea level. This movement of water from polar regions causes some minuscule yet measurable changes; the flow of meltwater alters gravitational fields, causes continents to rise, and even affects the Earth's rotational rate. Statistical studies also indicate that the amplitude of some areas' tides—that is, the mean of higher high water minus the mean of lower low water is slowly changing. The results of these changes are increased beach erosion or sediment depositions in some areas and other coastal transformations. Spring tides coinciding with storm surges will continue to displace millions of people, cause devastation, change national policies, and alter the course of history.

Tidal Bores

There are a number of places in the world where rivers enter the ocean via long, funnel-shaped bays. In such estuaries, especially during high spring tides, the broad front of the incoming tide wave is restricted by the narrowing channel and the shoaling water so that it abruptly increases in height and a visible wave front or *bore* exists. Beneath a tidal bore is the intense turbulent mixing of the water and the resuspension of sediments. Most bores are dull (except to the ardent wave researcher and the occasional surfer) and are regarded merely as a local curiosity, but, in a few places, they are respected or even feared. George Airy, one of the founders of wave theory, observed the bore in the River Severn, near Gloucester in England and wrote that he was thrilled by "the visible advancing front of that great solitary wave"—the tide.

The Severn tidal bore is formed as the rising tide is constrained
by the funnel-shaped Bristol Channel between England and Wales.
Jamie Cooper/SSPL via Getty Images

Actually, on entering shallow water, the solitary wave front often
breaks down into a series of small waves. The photo of the Severn
bore above shows a series of about six or eight short, steep waves
about 1 foot (0.3 m) high and 10 feet (93 m) long, moving up a small,
glassy-surfaced river. In other places the entry of the tidal water
causes the river surface to heave upward with an almost imperceptible
front; in the space of two minutes the water level rises by 3 feet (1 m)
or more.

A few famous bores have steep-breaking fronts that connect the
original water surface with a new surface at a substantially higher level.
The photo on page 145 shows the Qiantang River in northern China,
where there is a sudden contraction of the channel, the bore is more
than 16 feet high (5 m). It extends across the river more than 1 mile
(1.6 km) in width, and travels at about 12 knots.

The Qiantang River tidal bore rises several meters over the riverbank near Hangzhou Bay, China, on August 22, 2013. *VCG via Getty Images*

Visitors to the Hangzhou Bay waterfront have been amazed to see the local boatmen suddenly paddle frantically for the riverbank and apparently without cause pull their boats out of a placid stream. In a few minutes the breaking bore passes, the boats are returned to the river at its new, higher level, and work resumes.

The bore of the Amazon is even more spectacular—attaining a height of 25 feet (7.6 m). Seen from the high dikes near the river mouth, it has the appearance of a several-miles-long waterfall traveling upstream at a speed of 12 knots for 300 miles (480 km). The roar can be heard for 15 miles (25 km). Other tidal bores are found in many locations around the world including the Batang River in Malaysia, Styx River in Queensland, Australia, the Seine in France, and the Bay of Fundy in Canada, to mention a few.

Seiching

If the surface of an enclosed body of water such as a lake or bay is disturbed, long waves may be set up, which will rhythmically slosh back and forth as they reflect off opposite ends. These waves, called *seiches*, have a period that depends on the size and depth of the basin. They are a rather common phenomenon, but because the wave height is so low and the length so long, they are virtually invisible, and few laymen are aware of their existence. Seiches have been observed in the Greenland fjords after the glacial calving of icebergs; in the Adriatic, seiches contribute to the flooding of Venice, Italy; and winds across the Baltic Sea set up seiches that contribute to inundations of St. Petersburg near the Neva River delta.

Seiches can be regarded as standing-wave patterns or as the reflection of trapped waves. A pattern of standing waves (in contrast to the progressive waves of the open ocean) is composed of nodes, at which the height of the water surface stays the same, and *antinodes*, where the surface moves up and down. The nodes and antinodes maintain a fixed position, as do the surface particles of water, but beneath the surface there are swift currents as the water shifts to support the changing waveform.

Natural basins are so irregular in shape and depth that it is best to begin by thinking about the nice, clean situation that exists in a bathtub. The tub in figure 30 is nearly rectangular and has almost vertical frictionless sides—thus fulfilling the requirements of simple theory. If you put about 6 inches (15 cm) of water in your tub, you can make standing waves or seiches.

First set the mass of water rocking with a fundamental wave that reflects back and forth off each end. The midpoint of the tub is a node, and the water depth there will remain the same (6 inches, or 15 cm), whereas that at each end will vary from, say, 4 to 8 inches (10–20 cm). If you add some neutrally buoyant marker, such as small wads of paper that float at mid-depth and move with the water particles, the motion of the water in these standing waves becomes apparent. Beneath the nodes there are high horizontal velocities; at the extremities, the motion is mainly vertical. The period will be about two seconds. The fundamental period is that of a wave whose length is twice the distance between reflecting boundaries.

Now float a piece of board crossways at the midpoint and pump it up and down at a rate of about once a second. The water in the tub will now resonate in a different fashion; there will be two nodes and three antinodes. This is the *first harmonic*, which is twice the fundamental frequency.

These measured periods confirm the simple formula for the natural period for a closed basin with a depth d. $T_n = \dfrac{2L}{(n+1)\sqrt{gd}}$ in which L is the length of the tub (about 4 feet, or 1.2 m), $\sqrt{gd}$ is the velocity of a long wave; n is the "order" of motion (fundamental = 0, first harmonic = 1, etc.); g, of course, is gravity.

Thus, the natural period of the tub is two seconds, and its first harmonic is one second. For a harbor a mile (1.6 km) across, averaging

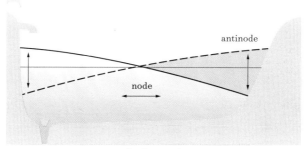

fundamental

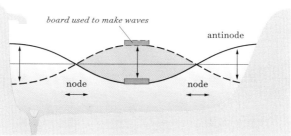

first harmonic

FIGURE 30: The term seiches describes the sloshing of waves back and forth in a bathtub, lake, bay, or enclosed or partially enclosed body of water. Seiches have atmospheric, surface wave, or seismic (wave) origins.

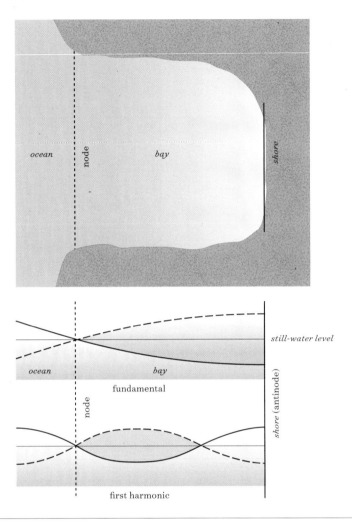

FIGURE 31: Seiching in an open-ended bay.

50 feet deep (15 m), the fundamental period is 264 seconds, and the first harmonic is 132 seconds. Other, higher harmonic orders may also be present at the same time.

Since few harbors have the neat boundaries of a bathtub, the waves do not reflect evenly; therefore, it might be expected that long-wave patterns would be difficult to establish. Not so. Even harbors with exceedingly irregular shapes seem to rock with remarkable regularity, which is recorded on the tide gauges. Moreover, many harbors and bays have a large part of one side open to the ocean. This has two major effects: (1) the laws governing the seiching are different; and (2) disturbances from the ocean can easily enter and start the harbor seiching.

If one side of the bay is open to the ocean, the missing reflecting surface is replaced by a nodal line, as shown in figure 31. Then, the fundamental mode of oscillation is that in which the opening is at the first node and the first harmonic has the second node at the opening.

Tidal recordings made on open coasts also show seiche-like motions of the water surface that can be interpreted only as oscillations of the water masses on the continental shelf. These are believed to exist because the shelf acts like an open bay. Thus, we discover that neither the bathtub nor the bay is necessary for this kind of wave to exist.

The Wave of No Return

We approached the island by open boat, slowed, dropped the anchor, and the boat came to rest. The Mediterranean island of Tinetto is just outside the safe harbor of La Spezia, Italy, and exposed to the deep-water swell of the Ligurian Sea. The island is small; it can be hiked end-to-end in a few minutes, and it rises only a few stories above the sea. The water was clear and the warm, gentle breeze warmed our skin as we waded across the shoal to the shore.

When we were upon Tinetto, the island revealed an unusual feature: an opening filled with salt water—an oblong well just wide enough to stretch out one's arms and long enough to lie in prone with extended dive fins. Intrigued by this well-like opening, which seiched up and down with the rhythm of long-period waves, I descended, with mask and fins, into a pulsing void below. Facing downward, my eyes slowly adjusted to the dim light. With a deep breath through my snorkel, I bent at the waist, raised my fins in the air, and sunk headfirst. After a distance of ten body lengths and thirty seconds below the surface, a submarine cavern appeared with its own white sand beach, which continued gently downward toward a distant light. Each unseen passing wave stirred grains of sand back and forth in the sand ripples on the cave's floor.

After a minute on the bottom it was time to return to the surface. Back above water I exclaimed to my companions, "There's a sandy beach down there and I can see light coming from the other side. This cave must extend all the way through Tinetto. It's an underwater tunnel."

I dove again and again, each time deeper and longer. The submarine beach ended at a restricted opening the size of a doorway, where the oscillating flow had scoured away all the sand. Here I guessed would be my point of no return on my next dive. Back at the surface, I told my companions I would go for it on my next dive. "If I don't come back in three minutes, don't follow me. Wait until I swim around the island."

I calmed myself and dove again, this time intent on swimming past the doorway and through the cave entirely. Below, each wave's outward rush of water sucked me toward the doorway. I swam further and then paused at the threshold. Tranquil, yet focused, I passed the point where I was too far into the cave to make it back the way I had come. The surge of the next deep-water wave's energy was even stronger; this was the wave of no return, and forward was the only way out. As I continued downward, I equalized the pressure on my eardrums. When the next wave forced water against me, I resisted by holding onto the rocks on the bottom. I had pushed harder and deeper than I had anticipated, now the equivalent of six stories underwater—holding my breath with no point of reference, I realized I was far from safety.

After about two minutes I cleared the sloshing narrows into the beckoning light and rotated upward, expecting to ascend a vertical wall to the surface. But there was more—above me was an unseen wave-worn roof to the cave which extended upward, not vertically but at a 45-degree angle. I briefly thought of the possibility of a shallow-water blackout, but pushed the thought out of my mind. I passed half a football field in length before reaching the surface. In the span of just over two minutes, I had swum underwater through an island, felt the pulse of wave energy, resisted the movements of water and sand, and almost passed out. The sun warmed my lips, and I began to smile, now unchallenged by the sway of the seiche. – KM

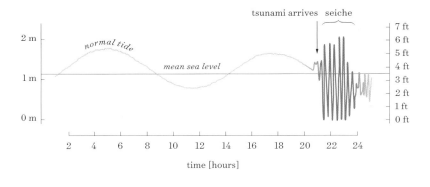

FIGURE 32: Seiching in Pago Pago, Samoa, excited by the tsunami of May 22, 1960.

CAUSES OF SEICHING

The question now arises as to the source of the disturbance that causes seiching. In a lake or other completely enclosed basin a sudden change in atmospheric pressure, such as would be caused by a wind squall passing over one end, is the most likely cause. But seiching in bays that open into the ocean is almost invariably caused by the arrival of a long-period wave train. A Pacific tsunami will usually succeed in exciting all the bays and harbors around its rim. Often these will oscillate for days, producing tide-gauge records similar to that shown in figure 32 of Pago Pago, Samoa, on May 22, 1960.

After the water is set in motion by the first arriving wave, seiching at the natural period of the harbor is likely to mask subsequent wave arrivals, thus making it difficult to obtain the period of the tsunami itself from the tide gauge. At Pago Pago, for example, the natural period of the harbor is twenty-two minutes, and its precise rhythmic motion is quite unlike that of other harbors disturbed by the same tsunami. If, by coincidence, the period of the tsunami is an even multiple of the natural period of the harbor, then the seiching motion is amplified by each new wave that arrives, and the water motion inside the harbor may become more violent than the motion outside. The fact that tsunamis are relatively rare suggests that long-period waves from other causes are responsible for creating most seiches.

Harbors and Seiches

The phenomenon of seiching is rarely troublesome to civilization. Exceptions being in some harbors where moored ships are moved about by these long, low waves and in marginal seas as mentioned in the Baltic and the Adriatic—this is known as surging. For long waves of the same height (these waves are usually less than 1 foot high, or 0.3 m) the amount of horizontal water motion is in proportion to the period. Thus, seiches with a period of several minutes can cause large ships tied to piers to strain at their lines and small ones at anchor to perform strange gyrations.

It is not easy for an observer to decide just what effect these waves have on ships or other floating objects, because the motion is slow. In 1946, while studying surging in Monterey Harbor, time-lapse pictures of the fishing fleet at anchor in the bay were taken. The boats were tethered to moving buoys by a single bowline and thus were reasonably free to move. A camera was set up on the bluff over the harbor to take pictures of the boats at the rate of one frame per second. When these images were replayed at sixteen frames per second, the surging motion emerged. Clear patterns of water motion became evident as the previously indiscernible three-minute surge was compressed into eleven seconds. The relatively quiet Noyo River and Santa Barbara Harbors were similarly photographed on days when no surging motion could be observed. To our surprise, a substantial surge could be seen. The boats would strain on their lines in unison, first in one direction and then in the other, with remarkable regularity.

In most harbors the surging is more a scientific curiosity than a serious problem, but Los Angeles harbor in California is an exception. There, even when the water surface appears calm, large ships tied up may move as much as 10 feet (3 m), snapping heavy mooring lines, breaking piles, and damaging the ships themselves. These invisible surges have periods of three, six, and twelve minutes, corresponding to the natural periods of the basin. Because of seiches, even a small tsunami, with the right period, can cause damage to boats in a well-protected harbor.

Monterey Harbor

In the 1980s, I was involved in a study to build a small marina inside Monterey Harbor, and the seiching remained a concern. Today, adjacent to the Coast Guard pier in Monterey Harbor, California, you can still watch the effects of the seiche as the boats surge back and forth in the harbor just as Willard Bascom did over seventy years ago. Seiches abound. Go and see. – KM

TIDAL POWER

The rise and fall of tides have been used for power since antiquity. The principal requirement is a large bay with a narrow entrance where the tidal range is large. It is the long-sustained rush of water through channels that makes tidal power possible. Not many places in the world have the required embayment and the large tides. The Woodbridge Tide Mill in Suffolk, England, has existed at least since 1170 AD and is still in operation. Hundreds of other tide mills have existed through the centuries in many regions in France, the Netherlands, and in North America, including Slade's Mill for grinding spices at Chelsea, Massachusetts, built in the 1700s.

The tidal plant at the entrance to La Rance estuary in Brittany, France, was first connected to the French electrical power grid in 1966 and is still in operation. The tidal range there reaches a maximum of 44 feet (13 m), and the area of the bay is about 9 square miles (23 km²). The plant turbines are more than 17 feet (5.35 m) in diameter and can generate up to 240 megawatts of power.

The largest existing power plant is the Sihwa Lake Tidal Power Station in South Korea. Additional smaller power plants exist in Canada, China, the Netherlands, Russia, and the United Kingdom, with several large (more than 1,000 megawatts each) power plants in the proposal stages.

The British government estimates that more than 7,000 megawatts of energy could be extracted from tidal energy in the United Kingdom. In 2003, the House of Commons Science and Technology Committee created the European Marine Energy Centre (EMEC) Ltd. and established a wave and tidal power research center in Orkney, Scotland.

As of January 2018, the Scotrenewables Tidal Power SR2000 is operational with a nominal 2-megawatt grid-tied system that supplies 7 percent of the Orkney electrical power requirements. This system is referred to as a *tidal stream* because it harvests the flowing tidal currents rather than using a barrier as found in La Rance Bay in France.

The Japanese have a national effort to convert tidal or open ocean currents into usable power. Toshiba Corporation was selected in 2014 for research and development of a turbine system driven by the ocean currents. In 2017, IHI Corporation unveiled an ocean current power generator to be deployed along the coastline from Kamaishi (Iwate Prefecture) in the north to Kumejima (Okinawa Prefecture) in the south. Several ocean current energy systems have been successfully used in the English Channel, and additional plans exist for installations in New York's East River and a tidal turbine at San Francisco's Golden Gate. Extracting energy from the tides is a still a compelling area for entrepreneurial efforts.

TIDAL STREAMS Tidal streams are swift currents (usually tidal and in open water) that can be utilized as a source of renewable energy. This is a composite image of a turbine in the waters off Eday Island, Scotland. *Five Square Imagery/ANDRITZ HYDRO Tidal Turbine*

7   Impulsively Generated Waves

When energy is released, nature likes to form waves. And when a force is suddenly applied to a water surface, waves are generated. The impulse of a pebble tossed into a pool sends out a series of waves in concentric circles. In the ocean, an earthquake, a volcanic eruption, a landslide, or a nuclear explosion may produce the same effect on a much larger scale. Large events of impulsively released energy will radiate through the air, land, and water for thousands of miles. The train of waves leaving such an event often contains a huge amount of energy and moves at high speed. As a result, the waves may be tremendously destructive when they encounter a populated shore.

The general public has occasionally referred to these waves as tidal waves, much to the annoyance of oceanographers, who are acutely aware that these waves have no connection with the tides. In an effort to straighten out the matter, they adopted the Japanese word tsunami, which now is in general use. As previously mentioned, tsunami means

PREVIOUS SPREAD: A sudden underwater release of energy off the Tōhoku region of Honshu, Japan, caused a tsunami and changed the lives of millions of people in March 2011. *Kyodo News Stills via Getty Images* "harbor wave" in Japanese, which also isn't exactly accurate. Either way, semantics aside, tsunami is the term we will continue to use in this book.

SEISMIC SEA WAVES

A somewhat more descriptive term that applies to waves caused by earthquakes is seismic sea waves. There are several mechanisms by which earthquakes can generate seismic sea waves, two of which are illustrated. The first (see figure 33, page 160) is a simple fault in which tension in the submarine crustal rock is relieved by the abrupt rupturing of the rock along an inclined plane. When such a fault occurs, a large mass of rock drops rapidly and the support is removed from a column of water that extends to the surface. The water surface oscillates up and down as it seeks to return to mean sea level, and a series of waves is sent out. If the rock fails in compression, the mass of rock on one side rides up over that on the other, and a column of water is lifted, but the result is the same: a tsunami. Tsunami waves are created when sea level is disturbed as energy is released. And the bigger the fault, the more energy available to form waves.

The amount of energy released in a seismic event is related to the Richter magnitude scale. In December of 2004, a massive underwater earthquake off the west coast of Sumatra (measured 9.1 Richter) was the cause of extensive destruction throughout the Indian Ocean. The Sumatra earthquake increased the global awareness of the destructive forces of a tsunami and will be discussed later in this chapter. Figure 33 (page 160) shows the mechanism of an underwater earthquake as the source of a tsunami.

A second mechanism is a landslide, which is set in motion by an earthquake. If the slide begins above water, abruptly dumping a mass of rock into the sea, waves are made by the same action as the plunger in the wave channel. If the slide occurs well below the surface, it creates waves, as shown in figure 34 (page 161). To make the point, both drawings are much exaggerated. Actually, the water surface may fall only a few feet in water many thousands of feet deep, but that would not show at this scale. Off the west coast of Africa, there is evidence for and concern among experts about the possibility for a massive landslide

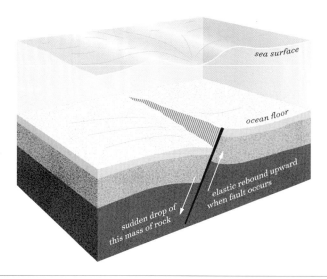

FIGURE 33: Tsunamis can be produced by a fault in the crustal rock that causes part of the sea floor to drop rapidly. The water surface above also falls, and a series of seismic sea waves is generated.

of the steep volcanic slopes of the Canary Islands, which could be large enough to affect the major shorelines of the Atlantic Ocean.

The waves created by landslides are very long and very low. Their period is of the order of a thousand seconds; their wave length may be as much as 150 miles (240 km); their height, only 1 or 2 feet (0.3 to 0.6 m) in deep water. The slope of the wave front is imperceptible, and ships at sea are unaware of their passage.

Because the wave lengths are so long, tsunamis move as shallow-water waves, even in the deepest ocean. As such, their velocity (C) is controlled by the depth.

$$C = \sqrt{gd}$$

Thus, if g = 32 and d = 15,000 feet or 4,500 m (the average depth of the Pacific Basin), the velocity of a wave in the deep Pacific is 692 feet per second (210 m/sec) or 472 mph (760 km/h). Fortunately, the Pacific is large enough so that waves moving at even that speed take considerable time to cross it, and a seismic sea wave warning system has been established to warn coastal inhabitants of approaching tsunamis.

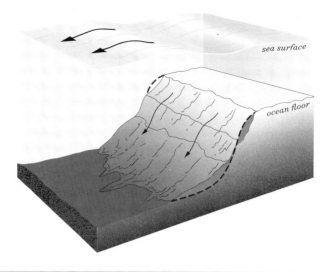

FIGURE 34: Another trigger for a tsunami can come from a landslide of loose sediment accumulated on the brow of a steep, submarine slope when it is set in motion by a nearby earthquake.

The events that cause a tsunami usually take place out of sight of land where they can do little harm. It is when these waves approach a coastline that they are at their worst. There, the influence of the bottom topography and the configuration of the coastline transform the low waves of deep water into rampaging monsters.

The first tsunami of which there is a record wiped out Amnisos, Crete, in about 1470 BC. One thousand years later, according to the ancient Greek, Pausanias, "The town of Helice perished under the waters of the Gulf of Corinth where the population was drowned to a man." In that millennium perhaps ten tsunamis were recorded. Now two or three a year cause localized catastrophes. Certainly there is no change in the activity of undersea earthquakes; the reason for the apparent increase is mainly that the world's population has grown so that people and wealth are now spread among once-deserted shores. Unfortunately, because this trend is certain to continue, the dangers to human activities from great sea waves are increasing. However, most municipalities, driven by permit fees, employment, and property tax revenues, continue to downplay tsunami risks and allow developers to build in low-lying areas.

One of the first efforts to catalog seismic sea waves was compiled by N. H. Heck of the US Coast and Geodetic Survey in 1947. In reading Heck's list of 270 events, one easily envisions great walls of water suddenly towering above frantic crowds, harbors being swept clear of ships, and soaked and terrorized survivors of the first wave racing the next wave to high ground. Table 7.1 contains some notable seismic wave events.

TABLE 7.1 Some Great Seismic Sea Waves

Date	Location and Description
Approx. 1650 BC	Thera (Santorini), Greece. Devastated major portions of eastern Mediterranean Sea, at the decline of the Minoan civilization.
July, 551	Beirut, Lebanon. Earthquake more than 7.0 on the Richter scale, ancient account of 30,000 killed.
September 14, 1509	Turkey. Sea came over the walls of Constantinople and Galata following earthquake.
December 16, 1575	Chile. Intense wave in the inner port of Valdivia. Two Spanish galleons wrecked.
March 25, 1751	Chile. City of Concepción was extensively damaged for the fourth time in a century by earthquakes. Sea withdrew and returned at great height several times. Disastrous effects at Juan Fernández Islands.
November 1, 1755	Portugal. Great Lisbon earthquake. Waves 15 to 40 feet high along Spanish and Portuguese coasts. Very high at Cadiz, where 18 waves rolled in.
December 29, 1820	Makassar, Celebes (Sulawesi), Indonesia. A wall of water 60 to 80 feet high swept over the fort of Boelekomba. Great damage at Nipa and Serang. A similar great wave at Bima, Sumbawa, carried ships over houses.
August 13, 1868	South Peru (now north Chile). USS *Wateree* was carried a quarter-mile inshore by a wave with a maximum height of 70 feet. The receding wave uncovered Bay of Iquique to a depth of 24 feet and then returned with a 40-foot wave, covering the city of Iquique.
August 25, 1883	Krakatau, Indonesia. Seventy percent of island disappeared, approximately 30,000 dead.
June 15, 1896	Northeast Japan. Sea waves nearly 100 feet high at head of bay and 10 to 80 feet elsewhere, 27,000 lives lost along 200 miles of coast, 10,000 houses swept away.
December 28, 1908	Messina and Reggio Calabria, Italy. Earthquake, damaged 90 percent of buildings resulting in a firestorm, more than 80,000 people killed, and caused flooding in Malta.
November 21, 1927	Aysén River region, Chile. Sea invaded land along 25 miles of coast. Vessels and crew flung into treetops of forest.
November 18, 1929	Burin Peninsula, Newfoundland. A tsunami from the Grand Banks earthquake swept up several narrow inlets to a height of 50 feet, destroying villages and causing major loss.
December 24, 2004	Sumatra, Indian Ocean. Worst tsunami in recorded history with more than 200,000 dead and $15 billion USD in damage.
March 11, 2011	Japan. Tōhoku earthquake and tsunami, 20,000 dead, Fukushima Nuclear Power Plant destroyed, over $300 billion USD in damage.

My own interest in tsunamis began on April 1, 1946; our field party returned to the Berkeley campus of the University of California after five months of daily observation of the waves and beaches of the north Pacific coast and we were greeted by someone asking us: "Did you see the big wave?" It sounded much like an April Fool's Day joke, but, sadly, it was not. There had been a landslide in the Aleutian Trench early that morning and its waves were wreaking havoc around the Pacific Basin. After the bad luck of missing the actual arrival of the waves by a day, we set about collecting whatever data we could.

The next few days were spent questioning people who had seen the wave, surveying high-water marks, and photographing wrecked houses and stranded boats. Some of the stories were compelling, and each contained some useful fact that could be applied to the understanding of seismic sea waves. For example, we found that the first arriving crest is often so small it is unnoticed, but it is soon followed by a major recession of the water.

The arrival of the trough of one of these great waves should serve as a warning, but instead it attracts the curious, who often follow the receding water out to pick up flopping fish and look at the newly exposed bottom, instead of running for high ground. When the next crest arrives, it might come quickly—in some cases it can be a huge breaking wave—and the curious pay for their folly. This drop in water level over a period of several minutes, without a change in the appearance of the usual waves, is something like the rapid ebbing of the tide. In a similar way, the incoming crests may be seen only as a rapid rise in the general level of the water without any observable wave front. Doubtless, this tide-like action, which occurs in a few minutes instead of twelve hours, is partly responsible for the usual misnomer "tidal wave."

On the same occasion in the cove at Pacific Grove, California, a man was dozing on a bench 15 feet (4.5 m) above the normal water level. He awakened when one dangling hand was made wet by the gently rising water; he sat bolt upright on the still-dry bench to watch the surrounding water slowly recede again. At the same instant, in nearby Monterey harbor there were unusual currents but no important rise or fall of water level. These incidents raised an interesting question. Why should there be this major difference in the height of the wave at two points

IMPULSIVELY GENERATED WAVES   **163**

The tsunami of May 23, 1960, arrived as a wave of translation at the Hilo waterfront, Hawai'i, and deposited wooden houses like driftwood at the high-tide mark. *Pacific Tsunami Museum*

only a mile apart? Part of the answer seems to be that Pacific Grove faced away from the wave, whereas Monterey faced into it.

Later, over a period of years, I traveled to many Pacific shores asking about the effects of that 1946 tsunami. Remarkably, points facing into the waves and bays facing away from them were often the hardest hit. For example, Taiohai village at the head of a narrow south-facing bay in the Marquesas Islands 4,000 miles (6,400 km) from the earthquake epicenter was demolished. Hilo, Hawai'i, only half as far from the disturbance and whose offshore topography seems precisely suited to funnel tsunamis toward the town, fared worse. One hundred and seventy-three people died, and millions of dollars in property damage were caused by the waves at Hilo that morning.

But the truly great waves of April 1, 1946, struck at Scotch Cap, Alaska, only a few hundred miles from the tsunami's source, where five men were on duty in a lighthouse that marked Unimak Pass. The

The remains of the Scotch Cap Lighthouse on Unimak Island, Aleutian Islands, Alaska. The tsunami of April 1, 1946, swept away the lighthouse and killed the five-man crew—the worst disaster to happen to a land-based Coast Guard light station. *US Coast Guard/Alaska State Library Historical Collection*

lighthouse building was a substantial two-story reinforced concrete structure with its foundation 32 feet (10 m) above mean sea level. None of the men survived to tell the story, but a breaking wave over 100 feet high (30 m) must have demolished the building at about 2:40 am. The next day, Coast Guard aircraft investigating the loss of radio contact were astonished to discover only a trace of the lighthouse foundation. Nearby, a small block of concrete 103 feet (31 m) above the water had been wiped clean of the radio tower it once supported.

Tsunami Warning Systems

Largely as a result of the Hilo disaster, a seismic sea wave warning system was developed by the US Coast and Geodetic Survey. Today the National Oceanic and Atmospheric Administration's Pacific Marine Environmental Laboratory in Seattle, Washington, continues this work in conjunction with an international network of seismic alerts. It works like this: Many seismograph and pressure sensor stations around the Pacific Rim and from Peru to Japan are equipped with automatic alarm systems and recorders. Dozens more around the world augment the system. When the tremors from a large earthquake are received, the alarm sounds, alerting the local observers, and the recorded data is reported to a central station. If the analysis shows that the quake is located under the ocean (using the arrival times of the first earthquake shock at the various stations), a message is sent to tide-measuring stations nearest the quake's epicenter. This message contains the estimated times of arrival of a possible tsunami. Each station is asked to report back whether such waves actually arrive. If unusual wave activity is reported, a warning is issued to authorities in coastal areas that may be affected. Attempts are made to predict the height of the waves to be expected, and then the actual measurements are compared to the predictions.

The tsunami warning system continues to evolve, and all the major ocean basins are now covered, including the Mediterranean. The Comprehensive Nuclear-Test-Ban Treaty Organization (CTBTO) has a global array of seismic and hydroacoustic stations to detect nuclear explosions. The organization, which is a United Nations entity, has agreements to provide many nations around the globe with tsunami-related information from approximately 100 CTBTO stations.

In 1950, a warning instrument was mounted at the outer end of the Hilo pier. It was designed to detect waves in the 1,000-second band midway between the longer-period tides and the shorter-period swell. The drop in sea level, which precedes the arrival of a tsunami, activated an alarm bell in the police station some distance inland and the city was warned in time to flee to higher ground. Today these systems are electronic and computerized, and with smartphone applications for individuals.

The need for Pacific Rim cities to have a wave warning system was clear enough, and over a period of years, a combination of false alarms and small tsunamis made it possible to "work the bugs out of the system." When needed for one of the greatest tsunamis of the past century, it was ready.

On May 22, 1960, a violent earthquake (magnitude 8.5 on the Richter scale) shook the coast of Chile. A volcano erupted; there was widespread faulting, subsidence, and hundreds of landslides. In a local disaster area 500 miles (800 km) long, 4,000 people died, half a million homes were damaged, and $400 million USD worth of property was destroyed. There was also a major subsidence on the great undersea fault that parallels the coast, and this generated a tsunami that was felt on all Pacific shores.

In Chile, dozens of waterfront towns were devastated. Coastal cities in New Zealand, Australia, and the Philippines were flooded by several feet of water. On the US coast, Los Angeles and San Diego harbors in California suffered damage to piers and small craft. In Japan, 9,000 miles (14,500 km) from the origin, the waves were as much as 15 feet (4.5 m) high. There, 180 people died, and the damage was estimated at $50 million USD. But Hilo, Hawai'i, again took the worst blow; although the property damage was more serious than in 1946, this time the population was warned and, thankfully, there were few deaths.

The Tsunami Warning Service has grown considerably and is now a cooperative international organization. With an expanded network of stations and decades of experience, the reliability of its forecasts has improved. Fortunately, most tsunami watches and many warnings are not followed by destructive tsunamis. Newer methods have improved tsunami predictions; the use of tide gauges has been combined with the information from many seismograph stations. To predict whether a tsunami has been formed, details of the location and magnitude of the earthquake must be known. As a general rule, destructive tsunamis are generated by earthquakes of 8 or stronger on the Richter scale, depending somewhat on the exact location of the quake. Therefore, any Pacific earthquake greater than 7.5 initiates a tsunami watch, but an official warning is issued only after tide gauges detect a wave.

One problem is that the magnitude estimates now come from seismographs that are particularly sensitive to seismic waves with periods of fewer than twenty seconds. Longer waves, which contain much of the energy, are excluded. This means the quake location is pinpointed by means of the higher-frequency seismic waves, but the earthquake amplitudes are derived from the intermediate length (twenty-second) waves. Some seismic researchers think that twenty-second waves do not accurately reflect the true magnitude of some earthquakes and that 100-second waves should be used instead. Otherwise, much of the energy released by large earthquakes, which might make tsunamis, goes undetected. Additional tools are now available to help researchers predict the arrival times and magnitudes of tsunamis. The massive Sumatran-generated tsunami (2004) was even detectable in satellite data, including that from the Jason-1, TOPEX/Poseidon, and Envisat satellites.

The Chilean earthquake of 1960 had a magnitude of 8.3 when calculated on the basis of twenty-second waves, but its magnitude was 9.5 if the longer waves were included. The difference represents ten times greater wave amplitude and sixty times more energy released. The long-period waves detected by seismographs are better indicators of the amount of energy available for tsunami generation.

The remarkable thing about the ocean is how calm and stable its surface can be. Considering its breadth and depth, the changes in height caused by waves and tides are insignificant, except to those who live at the water's edge.

A fascinating example of a seismic sea wave generated by an above-water landslide is the Lituya Bay incident. Lituya Bay, on the Alaskan coast, is an active earthquake region. Two glaciers flow into the upper end of the steep-sided bay and near its center is Cenotaph Island. A sandspit across the mouth keeps out the big waves from the Gulf of Alaska so that fishermen regard Anchorage Cove, just inside the entrance, as a safe haven.

On July 9, 1958, two fishing boats, the *Badger* and the *Sunmore*, were anchored just inside the spit when a major earthquake occurred. The shock started landslides that cleaned the soil and timber off the mountainsides at the upper end of the bay at 1,800 feet (550 m) above sea level, and at the same time caused great masses of ice to fall from the front of the glaciers into the water.

An eyewitness account by the skipper of the 40-foot (12-m) *Badger* gives a vivid picture of the result. He felt the earthquake, and, looking inland, saw the first wave building at the head of the bay. As it passed Cenotaph Island, he estimated its height at 50 feet (15 m)

The December 2004 earthquake in Sumatra, Indonesia, generated a tsunami that devastated coastal areas in Asia and Africa. It killed more than 200,000 people. The province of Aceh lies only 60 miles (100 km) from the earthquake's epicenter and sustained what many consider the worst tsunami damage. *Guy Gelfenbaum, USGS Pacific Coastal and Marine Science Center*

(measurements made later indicated it probably was much higher). It swept through Anchorage Cove, carrying the *Badger* over the spit at an altitude of about 100 feet (30 m) and dropping it into the open sea. There the boat foundered, but the boatman was able to launch a skiff, and he and his wife were picked up by another fishing boat. The 55-foot (17-m) *Sunmore* was not so fortunate; it was swept against a cliff, and no trace of it or its crew was ever found. This wave, although very high, was very localized. The somewhat unusual circumstance of its origin leads one to muse on the great waves that must have been generated in billions of years of geological time. One can visualize the walls of water that must have raced outward when whole mountains suddenly slid into the sea, or when a continental perimeter abruptly shifted, or when a great meteor landed in the sea—a pebble in Earth's puddle.

We could not have imagined the extent of devastation caused by the massive Sumatran earthquake on December 26, 2004. It had no parallel in recorded history—an estimated 200,000 humans perished in Asia and Africa, and millions were left homeless. The shores of many countries around the Indian Ocean experienced major destruction due to the tsunami that reached over 33 feet (10 m) in height in many areas. In Sri Lanka, 600 miles (1,000 km) away from the earthquake, a train was swept off the tracks, taking as many as 1,000 passengers to their deaths. Most of Sri Lanka's road system in the south was washed away. Throughout Southeast Asia, thousands of vacationing tourists were killed. The Maldives, with 300,000 people on 200 islands, had several low-lying islands washed away. Indonesia experienced extensive transport and communications collapse, delaying relief efforts, and in Phuket, on Thailand's southern coast, thousands of foreign tourists died at popular seaside vacation spots. In Africa, at least 3,000 miles (5,000 km) away, more than 100 fishing vessels were lost off the coast of Somalia. The energy from the tsunami propagated to all the world's oceans and is estimated by the US Geological Survey to be 2×10^{18} Joules, or 475,000 kilotons (475 megatons) of TNT, or the equivalent of 23,000 Nagasaki bombs. Sea-level changes were measured as far away as San Diego, California. The physical devastation was followed by economic and societal turmoil. Fifteen years later, life had not yet returned to normal—it never will.

In March 2011, a magnitude of 9.0 earthquake occurred 80 miles (130 km) east of the Tōhoku region of Honshu, Japan. It was the most powerful earthquake on record in Japan. The earthquake created a tsunami that devastated the eastern coast of northern Japan (photo on pages 156-57), in some areas it reached an estimated height of almost 130 feet (40 m), traveled up to 6 miles (10 km) inland, destroyed more than 100,000 buildings, and killed some 20,000 people. The tsunami waves were more than 40 feet (13 m) in height on arrival at Sendai in the Fukushima Prefecture. Because of the tsunami, the nuclear reactors in Fukushima failed catastrophically, leading to the release of large amounts of radiation (including cesium-137, with a half-life of around thirty years) into the Pacific Ocean and atmosphere. An estimated 160,000 people had to flee the area because of elevated radiation levels, and eight years later, in 2019, the radiation from the reactors in Fukushima was still leaking into the Pacific and was detectable in the ocean waters of Hawaiʻi, Southern California, and Mexico. The spread of contaminants continues. In October 2019, the floodwaters of Typhoon Hagibis washed away an unknown number of bags (weighing about 1 ton each) still laden with radioactive contaminants from the 2011 accident.

In 2012, scientists at Woods Hole Oceanographic Institution estimated that 16.2 petabecquerels (10^{15} becquerels) of radiation had been released in Fukushima. The Tōhoku tsunami caused the most costly natural disaster in human history, estimated at more than $300 billion USD. Since the Tōhoku tsunami, the Japanese government has established the Fukushima Renewable Energy Institute, a large project to promote research and development in renewable energy. The tsunami that struck Fukushima caused financial, societal, and geopolitical changes around the world. One result of the accident was in Germany where political activism caused the shutdown of eight of Germany's seventeen nuclear reactors. The shutdowns are part of Germany's "Energiewende," a transition to low-carbon emission, renewable energy. Since 2011, Germany has invested billions of euros in renewable energy infrastructure, including many large offshore windfarms. These renewable energy sources are at times so abundant that Germany can sell excess capacity. Germany intends to shut down all of its nuclear reactors by 2022, which has caused international energy companies (from Russia, Norway, the Netherlands, and others) to compete for Germany's natural gas needs. Such energy disputes frequently lead to wars.

EXPLOSION-GENERATED WAVES

Tsunamis may also be generated by large, violent explosions, which are fortunately less likely than an undersea earthquake. The classic example of an explosion-generated wave is the eruption of the volcano Krakatau in the Sunda Strait, Indonesia. This story is best told by excerpts from the original report in the *Honolulu Daily Bulletin* on August 29, 1883:

> Krakatoa (Krakatau) erupted with the most violent
> explosions of recorded history. The entire north portion

of the island was blown away and in place of ten square miles of land with an average elevation of 700 feet, there was formed a great depression with its bottom more than 900 feet below sea level. Apparently pent-up superheated vapor exploded and ruptured the throat of the volcano allowing cold ocean water to "freeze" a crust on the rising molten magma there. Then, as with a safety valve tied down, the pressure began to build up. On the morning of August 27, 1883, this crust let go.

Over four cubic miles of rock was blown away; the sea was covered with masses of pumice for miles around—in many places of such thickness that no vessel could force its way through. Two new islands rose in the strait, the lighthouses were swept away. A column of dust rose seventeen miles and spread out so that at Batavia, a hundred miles away, the sky was so dark that lamps had to be burned in the houses at midday.

Eventually the emitted dust and sulfur dioxide were distributed by stratospheric winds and influenced the weather over the entire planet. The sound of the principal explosion was heard 3,000 miles (5,000 km) away and the atmospheric shock wave reflected off itself at the antipodes of the Earth.

But the most damaging effect came from the waves that inundated the whole of the shores of Java and Sumatra, which border the Sundra Strait. Many villages were carried away, including those of Tyringin and Telok Betong. The town of Merak, at the head of a funnel-shaped bay, was struck by a wave estimated at more than 100 feet (30 m) high. More than 36,000 people were drowned, and many vessels were washed ashore, including the man-of-war *Berouw*, which was carried 1.8 miles (3 km) inland and left 30 feet (9 m) above sea level. How the wave was formed must ever remain, to a great extent, uncertain—whether by large pieces of the island falling into the sea, by a sudden submarine explosion, by the violent movement of the crust of the Earth underwater, or by the sudden rush of water into the cavity of the volcano when the side was blown out.

Anak Krakatau, Indonesia, collapsed in December 2018, decreasing by two-thirds in height, a reduction of more than 700 feet (200 m), and generating a tsunami that killed hundreds of people. *Nurul Hidayat/ Bisnis Indonesia via AP*

The waves radiated westward from the Sundra Strait into the Indian Ocean, around the Cape of Good Hope, and northward through the Atlantic. Tide gauges in harbors at South Africa (4,690 miles, or 7,500 km, from the source), at Cape Horn (7,820 miles, or 12,500 km), and Panama (11,470 miles, or 18,500 km) clearly traced the arrivals of a series of about a dozen waves, which had traveled at an average velocity of about 400 mph (640 km/h). The period of the waves taken from tide stations near the explosion is about two hours; at great distances it is closer to one hour. A similar decrease in period with distance is noted in the records of the May 1960 tsunami.

Krakatau did not rest—forty-five years after the 1883 eruption, a small island began to emerge from the sea, called Anak Krakatau or "Child of Krakatau." On December 22, 2018, part of the uninhabited Anak Krakatau island collapsed. It decreased more than 700 feet (200 m) in height, and as it slid into the sea it created a tsunami that killed more than 400 people and injured at least 7,000 inhabitants of the nearby islands.

An even larger eruption than Krakatau occurred long before, about 1600 BC on the Greek island of Santorini. The explosion and resulting tsunami were massive enough to alter the course of Mediterranean history. The center of the Minoan civilization in Crete, Knossos, was struck by the tsunami estimated to have been more than 330 feet (100 m) in height. Following the explosion, many civilizations weakened. Subsequently, the "Sea Peoples" descended along the Turkish coast, through Syria, and into Egypt. The periods of climate-induced droughts and famines played roles in the disruption of commerce. Political power changed hands and life in the eastern Mediterranean was never the same after the end of the Bronze Age.

Even further back in time, perhaps a million years ago, many cubic miles of the western side of the Hawaiian island of Oʻahu collapsed into the sea. The geological record and bathymetric surveys indicate that the resulting tsunami could have reached heights of over 1,000 feet (300 m), devastating any coastline in its path. Let us be thankful that today's civilizations had yet to arise. Some readers may think these references are too ancient to be relevant. However, the recent tsunami that destroyed the Fukushima nuclear power plant in 2011 not only disrupted world commerce but transformed the public's view of nuclear power and emphasized the possibilities of alternative energy. Nature's influence on politics cannot be denied; however, the politics of denial will not influence nature. Such patterns throughout recorded history can only shed valuable light on what's to come.

In early 1952, I was given the job of measuring the waves from "Ivy Mike," the first large thermonuclear explosion. It was to be exploded on the wide reef at the northern end of Eniwetok Atoll, in the Marshall Islands. Somehow it would make waves, but by what mechanism, and how large would the waves be? The best guide was the Krakatau report

The energy released in the Ivy Mike thermonuclear explosion in 1952 was small compared to an earthquake. However, there was great concern that part of the atoll could shear off in a landslide and propagate a tsunami across the Pacific. Eniwetak, Marshall Islands. *Courtesy of Los Alamos National Laboratory*

of the Royal Society of London, and I prepared an extensive abstract of it from which the preceding account has been taken. We guessed that the energy released by "Ivy Mike" would be about the same as that of the main explosion at Krakatau.

Six years earlier, in 1946, preparations were under way for the first underwater nuclear test inside Bikini lagoon. Many wave-generating experiments had been made in an attempt to forecast the size and period of the waves and the means by which they would be generated. Explosives were shot into ponds, and circular steel plates were dropped into basins that contained models of the entire atoll. There was a lot of splashing and an immense amount of data was taken, along with an abundance of theories. But even after photos of the actual waves made by the so-called "Bikini Baker" explosion had been studied, many uncertainties remained about how such waves were formed.

"Ivy Mike" was a different situation. It would be set off at sea level on a flat reef 1 mile (1.6 km) wide. On one side was the shallow lagoon, nowhere deeper than 200 feet (60 m). On the other side, the outer slope of the atoll dropped away into water over 12,000 feet (3,600 m) deep. We were nervously aware of the similarity of Krakatau, and we had to be sure that no large tsunami would be generated. If such a wave were released in deep water, it could cause damage all around the Pacific.

"Ivy Mike" would dig a large crater in the reef, but would it breach the outer edge of the reef? If so, the direct effect of the blast moving rock outward or the fast flow of water from the ocean back into the hole might start a wave. Experts were certain the crater would not be that large. Could the earthshock from the bomb start a landslide on the outer side of the atoll? Other experts were sure it could not because the outer slope was not quite steep enough. Could the rise in barometric pressure caused by the air shock start a wave? Yes, and although we could not estimate with certainty exactly how large it would be, we were sure that it would not be dangerous to people living around the Pacific.

The waves actually produced by "Ivy Mike" confirmed the advance opinions and although the hole in the reef was a mile in diameter and 600 feet deep (180 m), it did not breach the outer edge. A tsunami was generated with waves up to 20 feet (6 m), but it only affected the local islands; the amount of energy released was too small (equal to about

10.4 megatons of TNT) to generate an oceanwide tsunami. However, the radioactivity from 100+ subsequent tests persist, and some islands are still uninhabitable many decades later. Unfortunately, much more powerful weapons have been produced that release more radiation and could create a tsunami devastating the shores of an adversary. It is a terrifying image, and luckily since 1963 there have been a series of nuclear test ban treaties prohibiting underwater testing.

WAVES PRODUCED BY SHIPS

Any disturbance of the water surface creates waves, including the passage of a ship. But the waves made by ships are of a different kind and are studied for a different reason than we study the natural waves previously considered in these pages. Much of the power expended in propelling a ship goes into wave-making and anything that can be done to reduce these waves results in increased efficiency and is of direct economic and environmental importance. Consequently, some of the greatest names of hydrodynamics are associated with the ship-wave problem, including Bernoulli, Lord Kelvin, Rankine, and Froude. Today in the twenty-first century, scores of researchers continue this quest. Improved ship hydrodynamics are achieved using computational fluid dynamics and other methods to reduce the formation of ship waves, or wakes. Any reduction in fuel consumption, structural flexing, noise generation, or increase in propeller efficiency will reduce a ship's operational costs. Even the acoustic waves (sound waves) radiating from a ship is an unnecessary loss of energy. In commercial terms: energy is money, and escaping energy—in any waveform—is lost money.

A ship moving through the water is accompanied by at least three pressure disturbances on each side, which produce several trains of waves. In addition, the movement of the ship sets up an unusual traveling undulation that is not a wave in the ordinary sense, because it stays with the ship and is nonrepeating. This undulation is sometimes visible to a practiced eye if a ship is moving at slow speed through glassy water. It consists of two low mounds of water, one ahead of the bow, the other behind the stern, and a broad amidships depression. This special form of a standing wave seems to be an inescapable result of ship motion,

KELVIN WAVES The feathered "V" pattern diverging behind a moving vessel is created by the hull displacing water and is called Kelvin waves. A trail of stern waves is visible in the center. Santorini, Greece. *Kim McCoy*

because of the Bernoulli principle, that states that an increase in fluid velocity results in a decrease in the static pressure.

Lord Kelvin investigated analytically the pattern of waves generated by a pressure disturbance concentrated at a point and moving in a straight line. A thin stick drawn vertically through quiet water or even a small boat on a large expanse of smooth water will create such waves. This Kelvin wave system is characterized by: (1) *diverging waves*—a series of curved crests, concave outward and lying in echelon position; (2) *transverse waves*—convex forward and perpendicular to the direction of motion; and (3) a line of *crest intersections*, where the diverging and transverse waves meet, forming a constant angle with the direction of motion. These are shown in the photo above, in which the normally visible crest segments are indicated by heavy lines.

For large ships, whose hull would occupy a substantial part of the pattern just described, the mathematically ideal Kelvin system is replaced with arrays of pressure and velocity sensors. Various points along the hull generate waves, usually at changes in curvature along the

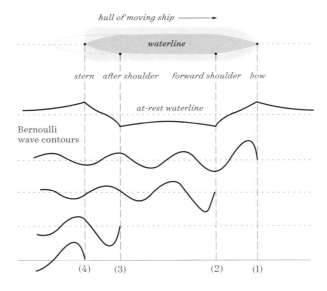

hull of moving ship ⟶

waterline

stern   after shoulder   forward shoulder   bow

at-rest waterline

Bernoulli
wave contours

(4)   (3)   (2)   (1)

FIGURE 35: The Bernoulli equation describes the relationship between speed and pressure. Water movement over the submerged contours of a moving vessel causes pressure changes, which results in a series of waves. These are complicated by varying ship speeds and oncoming ocean waves.

waterline—the wave-generating intensity related to the abruptness of change of direction. Four generating points, shown in figure 35, are the bow, the forward shoulder, the after shoulder, and the stern. Note that bow and stern create positive pressures, and the waves begin with a crest; at the shoulders, negative pressures create wave systems, beginning with a trough. When the vertical displacements of the water surface are added algebraically, the sum is an approximation of the resultant wave alongside the ship. Fortunately, the waves made by ship models in a towing tank are accurate predictions of those that will be created by the full-size ship. They represent and check the theoretical calculations and computational fluid dynamics (CFD) models so the designer should know what a ship will do in advance of actual construction.

These various sets of waves adjust their dimensions according to ship speed and consequently interfere with each other in a complicated way. As the ship's speed increases, the waves lengthen. Of course, the ship remains the same length, and each wave group still originates at the same point of curvature. But now the second and third crests in each system move farther aft; at some speed these will cancel, or reinforce, the waves of the following system. Thus, as a ship changes its speed, so too does the pattern of waves it makes.

A slightly submerged streamlined body, such as a submarine, at periscope depth, also generates surface waves. It creates a Bernoulli principle-based suite of transverse waves above its bow, midships, and stern points. Thus, even a submarine, if it is moving at a shallow depth, spends some of its power in making waves.

It is much less difficult to make waves than to avoid making them. A large part of a ship's power goes into unintentionally making waves, so a good deal of thought has been given to methods of reducing this wasted power. This is not the same as streamlining a ship to reduce its hydrodynamic drag. However well a ship is streamlined, its passage will still give the impulse that creates the various kinds of waves just discussed.

Of the waves produced by a ship, the bow wave is the largest, so attention first centered on how to reduce it. In the late nineteenth century, the British warship *Leviathan* was built with a large projecting underwater bow for ramming in the ancient Greek tradition. The ship's "superior performance" was attributed to this bow, and in 1907 David Taylor designed a similar bow for the USS *Delaware* in which the ram bow was enlarged and placed farther beneath the surface. Thus the bulbous bow was born. Its primary advantage is, of course, that the wave crest created by the movement of the bulb through the water coincides with the trough of the bow wave and the two mostly cancel each other out.

Subsequent warships were fitted with such bows, but without great success because the design rules were inadequately known. So, in the early 1930s, a ship model was towed without a bulb; then the bulb was towed without the model; until finally the model was towed with the bulb attached. Even though the bulb was crudely attached to the hull (not streamlined), there was a substantial reduction in resistance. This was developed by a British hydrodynamicist named W. C. S. Wigley into the basic theory for the bulbous bow. He found that at low speeds the total resistance of a ship increased because of the additional frictional and form drag of the bulb. At high speeds, the reduction in wave resistance (because of the interference between bulb and hull waves) overcomes the increased drag and there is a net reduction in total resistance.

A bulb bow on a ship can reduce drag by more than 10 percent.
At design speed, the bow creates a wave pattern that changes the
pressure distributions moving along the hull. *gtzx/Alamy*

Wigley concluded that: (1) the farther the bulb projects forward of
the stem, the greater the reduction in resistance; (2) the bulb should be
as low and wide as possible and still permit proper fairing into the hull;
(3) the top of the bulb should not be too near the water surface; and (4)
the useful speed range of a bulb is when the ship's velocity divided by
the square root of its length is from 0.8 to 1.9.

For a modern bulk cargo ship 900 feet (275 m) long moving at 30 feet
per second (9.1 m/sec) this comes out to be 1.0, and the bulbous bow
is useful. The ship's resistance is reduced 10 to 15 percent, and the
propulsive efficiency is increased 4 to 5 percent; thus, a reduction of
20 percent of the shaft horsepower in smooth water is possible. For the
ancient Greek trireme, 120 feet (36 m) long rowed at 9 feet (2.7 m) per
second, the corresponding number is 0.8—perhaps the ancients' use of
the "rostrum" was more than just as a battering ram.

Submarines can be tracked by means of their wakes. As they cleave
silently through the sea, they make waves, leave a trail of slightly

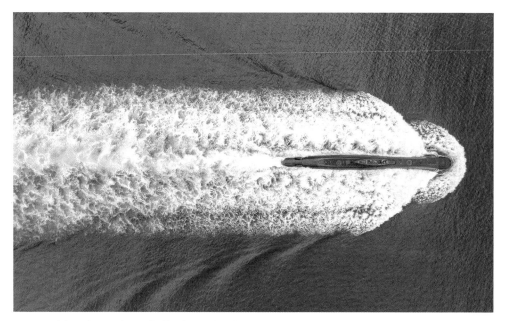

Submarines can be tracked by their wakes above and below the surface.
Johann Helgason/Alamy

warmer water, a trace of oil, or some other evidence behind. A great deal of effort has been spent on studies and experiments to detect these wakes. Aside from the obvious problem that there is a huge natural background of waves, temperature variations, and surface slicks, there are many other ships at sea that are also making waves and leaving other kinds of wakes. The image on page 183 shows one estimate of the distribution of the positions of merchant shipping underway on an average day. Unfortunately for submarine hunters, the number of large ships in the world's merchant fleet is estimated to be more than 50,000. The rotating propellers of a submarine mix up the water and leave a trail of turbulence behind. The lack of "normal" waves and normal ocean structure in the ocean is of great interest to submarine hunters.

Ships also may generate unseen internal waves on the interface between two layers of water of different density. In regions of melting ice or near the mouths of large rivers, a layer of fresh water often rests on the heavier oceanic salt water with little or no mixing. When this

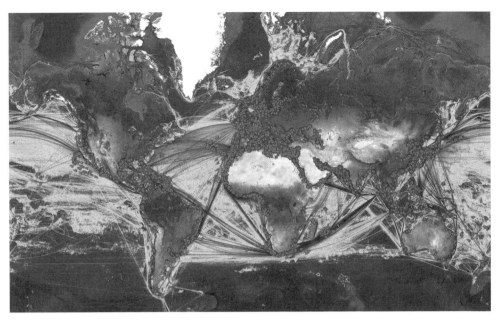

The automatic identification system (AIS) tracks merchant ships as they spin their impressive web of international commerce. More than 50,000 large ships of various types are underway at any given time, each traversing and creating waves. *Courtesy of MarineTraffic Density Map*

layering occurs, the progress of slow-moving ships is hindered because some of their propulsive energy goes into generating waves on the boundary between the fresh and salt water. These subsurface (internal) waves may be much higher and move much slower than the visible surface waves generated at the same time.

This phenomenon was studied by the Hall brothers in an Edinburgh wave tank in 1830. Much later, V. W. Eckman, investigating strange tales of Norwegian fishermen who claimed their boats got "stuck" in the "dead water" of fjords, gave the following explanation: the deep and still salt water of the fjord is "flooded" with a layer of fresh water. The bow of a fishing boat moving in the lighter-density upper layer causes a rise in pressure that depresses the fresh-salt interface just as though a thin, flexible membrane separated the two. This sets a train of waves in motion on the surface of the salt water, which moves at about one-eighth the speed of those on an air-water interface. These waves in effect "capture" the boat that creates them so that waves and boat move together

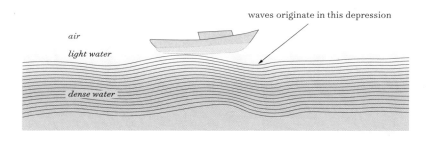

FIGURE 36: Slow-moving small craft generate waves on the interface between lighter and denser water.

as a unit. Then, the resistance of the slow internal waves adds to that of the ship. This is the reason why, once a ship is slowed down and caught in the position shown in figure 36, escape is difficult. Eckman suggested that fishing craft could avoid this difficulty by maintaining a speed above 5 knots.

Objects, including small craft and animals (such as porpoises, seals, and bodysurfers), can exploit the energy of the waves to propel themselves by sliding down the forward surface of an advancing wave. The object is thrust forward by a downwave force or slope drag as a vector connecting the gravity force to the buoyancy force, which always acts perpendicular to the still-water surface. When the slope drag is greater than the hydrodynamic drag (water resistance), the object moves at wave-crest speed. Today's surfing science provides an understanding about the complex dynamics when surfing the "sweet spot" of a wave, and is covered later in Chapter Nine.

The amphibious truck once used for surveying the surf zone is called a DUKW. DUKWs (pronounced "ducks") are fascinating amphibious six-wheeled trucks, 32 feet (9.7 m) long and 8 feet (2.4 m) wide, whose tops rise 10 feet (3 m) above the roadway. DUKWs can assume the proper slope and take advantage of an additional effect to surf on large breaking waves. Their front axles hang down in order to offer a vertical surface for the orbiting water particles to press against. Bodysurfers who hold their hands down beneath their bodies can get the same kind of boost. The air-water interface is a surface of constant pressure; beneath it are other parallel surfaces of constant pressure that move with imaginary waves that are subsurface reflections of the visible waves above.

Interestingly, porpoises are close to neutrally buoyant and have learned to tilt themselves at the proper slope to take advantage of the slope drag to surf on an underwater constant-pressure surface. These animals can ride beneath the bow wave of a ship indefinitely without appearing to exert any effort. Apparently a porpoise can do this because the skin drag of its outer body is less than the slope drag on the invisible surface.

It is even possible to surf on the waves made by a ship. As young boys on the Hudson River in New York, we used to paddle frantically to get a canoe into the proper position behind a ferryboat as it pulled away from the pier so we could get a free ride across the river, merely steering to hold position on the steep slope of the first transverse wave in its wake. It is also possible for boats to surf on their own waves. In the days when canal barges, drawn by horses on a towpath, were widely used for transportation, the horses soon discovered that if they temporarily sped up on approaching a narrow stretch of canal, they could then relax while the boat rode the waves of its own creation, or so reported Benjamin Franklin in 1768 after traveling on the canals of France. Many years later John Scott Russell studied "fly boats" on the Scottish canals where the same "advantageous principle was employed to reach high speeds in the passenger trade."

The canals were very shallow (probably less than 4 feet, or slightly more than 1 m deep) so that the waves moved at a speed of $\sqrt{gd}$, about 6 knots (7 mph, or 11 km/h). One can imagine that when the canal suddenly narrowed and the height of the bow wave increased, a wise horse (or driver) would smile at the prospect of surfing the load for a while.

8   Measuring and Making Waves

Scientific progress is largely dependent on the ability to make better measurements. Therefore, much wave research has been directed toward the development of new kinds of instruments and techniques for measuring waves. The understanding of how and why measurements are made gives one a much better insight into the nature of wave motion.

Until the early 1940s, direct observation, the photograph, and the tide gauge were the principal means available for studying waves. The observer would watch the sea surface and make notes. The number of seconds between wave crests (passing a piling or the bow of his ship) was recorded and an estimate of the height of each wave was made. The dynamic quality was lost in a series of uncertain tabulations: Wave crests cross each other, secondary crests ride on the flanks of large waves, and the crests a little way off are aligned with the trough being

PREVIOUS SPREAD: *FLIP* (Floating Laboratory Instrument Platform) in its vertical position, bow in the air, stern below. *FLIP* is a uniquely stable platform in this position, with only 15 percent of its slender 350-foot (107-m) length above water. Waves of all types—including surface, internal, and acoustic—have been measured from *FLIP*. *William C. Burgess*

An array of sensors measures currents, wave heights, sediments, and turbulence in the swash zone. Dr. Britt Raubenheimer adjusts the instruments in anticipation of a rising tide during the Nearshore Canyon Experiment in 2003. Black's Beach, California. *Marc Tule*

observed. Except near the shore, where the effect of the bottom tends to organize the waves, a thoroughly confusing situation exists, too complicated to be retained or analyzed by the human mind. Today, ocean waves, currents, and nearshore sediments are measured by using a variety of electronics exploiting light waves, radio waves, and sound (pressure) waves. From above, an array of orbiting satellites can measure waves, sediment plumes, coastal erosion-deposition, and primary productivity. These methods produce fabulous amounts of electronic data, all neatly digitized and fed into computer models.

WAVE OBSERVATIONS

The shape of the sea surface is "frozen" by photographing it. Single, still photos evolved into stereopairs taken from a ship's mast so that wave height could be obtained and the sea surface contoured. Today's surface-imaging techniques, with the addition of lasers, radio waves, and sound have exposed finer details over larger areas of coverage. Many of these methods have been adopted from land-based sensors and software, which helped create Google Earth. Imaging of waves is now possible under many conditions—at sea in storms, fast-motion of breaking waves on the beach, slow motion video of waves in tanks, and time-lapse movies of waves in harbors. Sometimes, in order to give scale to the photograph, markers are introduced; these include floating disks and spar buoys. A spar buoy is a slender, buoyant tube that floats vertically; often it is attached by a rubber cord to a horizontal damping disk that hangs far beneath. This reduces its vertical motion to a minimum as the waves pass.

Aerial photography for wave research introduced a whole series of possibilities. By analysis of precisely timed wave photographs, techniques were developed for measuring the depth of water in the approaches to the beach. In the 1940s, the Waves Project field party made synchronized photos in which pictures of the surf zone were taken simultaneously from an aircraft at 12,000 feet (3,600 m), the top of a cliff, and the beach. From analysis of these pictures it was possible to determine what waves were arriving and how they were affected by the underwater contours. The Sun's glitter (light reflection) pattern on the sea surface indicated the average slope, size, and direction of the waves to create a spectrum of the sea under various wind conditions. Aircraft in level flight at low altitude above the water can use a recording radio altimeter to get wave height. Submarines can point an echo sounder upward and record changes in the distance to the water surface above as waves pass. Or, in shallow water, a recording echo sounder in a small boat or personal watercraft will give an approximate image of the waves that raise and lower the vessel relative to the bottom.

Some methods were tools rather than instruments; they made wave observations more convenient and reduced the demands on estimation and memory. But a true instrument does something more. It amplifies sensory perception and makes it possible to learn things that could never be discovered by the ordinary human senses. Many of the present instrumentation had beginnings in the latter days of World War

II, when research was driven by the expected need to forecast wave conditions on enemy-held beaches where amphibious landings might have to be made. Today, with rising sea levels, new measurements help defend the land.

TABLE 8.1 Instruments for Measuring Waves

Property to be sensed	Means of sensing	How used
Light reflection	Camera, fixed and mobile	Object and pixel tracking, still and time lapse
Height of water surface	Float in pipe	Standard tide gauge
	Spar buoy	In deep water with deep damping disk
	Inertial measurement unit	Measures the heave motion of a ship
	Radio waves	Radio altimeters: aircraft and satellite
	Acoustic echo sounder	Pointed down from a pier in shallow water or up from a buoy, autonomous vehicle, or submarine
	Accelerometers	Anchored surface buoy "Waverider"
	Surface-piercing wires	Measures complex and rapidly changing wave surfaces
	Side-looking radar	Scans sea surface for roughness and height
Pressure at seafloor	Pressure sensor	The force per unit area is changed into an electrical signal
	Bourdon tube	Uncoiling tube drives a recorder
	Variable inductance	Measures change in a magnetic field
	Thermopile	Measures adiabatic heating of air
	Strain gauge	Measures change in length of metal
	Air bladder	Activates a recorder via air hose to surface
	Vibrator	Changes frequency as pressure changes
Water motion (velocity or acceleration)	Accelerometer	On buoy or ship to measure acceleration by waves
	Accelerometer, pressure combined	Wave recorder that computes wave height by using several sensors
	Rotor	Measures current caused by wave (in one axis)
	GPS drifter	Position vs. time, as it moves with current
	Acoustic Doppler	Measures "Doppler shift" caused by waves, currents, and the surface of the water
Drag	Strain gauge	Senses wave forces acting on an object
Impact	Dynamometer	Indicates maximum force or torque
	Diaphragm	Same as above plus hydrostatic force
	Piezoelectric disks	Electronic amplification of force

The development of an instrument begins with an analysis of the properties of the subject. What qualities of waves can be sensed and measured?

Table 8.1 (page 191) summarizes wave properties and lists instruments that have been used most generally. Although there are quite a few, the list is by no means complete. Wave investigators continue to devise new instruments and apply innovative methods to understand waves and coastal dynamics.

The simplest way to measure waves is to observe them passing a *wave staff*—a vertical pole on which a scale in feet or meters has been painted. While this book is not the place for a detailed description of wave-measuring devices, it is worthwhile to describe a few of the major forms of instruments that have contributed substantially to modern wave theory.

TIDE GAUGES

One of the first tidal gauges was invented by Lord Kelvin in 1882, and was further refined by the US Coast and Geodetic Survey. In the early designs, it was a pipe, perhaps a foot in diameter, open at both top and bottom and extended from near the harbor floor to well above the highest tide level. Inside the pipe (sometimes called a *stilling well*) was a float; from the top of the float a wire extended up around a driveshaft and down to a counterweight. A clockwork mechanism kept chart paper moving slowly beneath a pencil. As the tide rises and falls, it moves the float and pencil position back and forth, tracing out the height of the water on the paper.

When Lord Kelvin, best known for his abstract formulation of the second law of thermodynamics, presented his scientific findings with this instrument to the Institution of Civil Engineers, he was roundly criticized for having used a pencil instead of a fountain pen (which had just been invented). Responding to the derogatory comments, Kelvin said, "The ink marker has been tried for tide gauges and has hitherto been found unsuccessful on account of the slowness of the motion, but there is ample power in the tide gauge to drive a pencil." He further remarked, "Good workmanship is too often required to overcome the evils of a poor design." Today, most floats have been replaced by pressure sensors, the graph paper superseded by electronics, and long forgotten is the debate of fountain pen or pencil.

Tide gauges are usually set up in the quiet waters of a harbor, where they are not exposed to any swell, and their pipes extend deep enough so that the small waves generated inside the harbor do not affect the measurements. If the same device were attached to a pier extending out into the ocean, the measurements would reflect the rise and fall of each passing swell. This is because the open lower end of the pipe would permit the water to flow in and out rapidly, and the water surface inside would be at the same level as that outside.

However, a small change in the instrument converts it into a long-period wave recorder. If the bottom end of the pipe were completely sealed off, the water level inside would not change. But if this seal has a small hole in it, the pressure created by the passage of a wave crest will cause water to flow through the hole and raise the level inside the pipe slightly. Short-period waves and even ocean swell go by too quickly to permit enough water to flow through the hole to appreciably change the water level in the pipe. But long waves with periods of several minutes maintain the pressure long enough for the water level inside to respond. Therefore, even though these long waves are only a few inches high in the midst of a turbulent zone of waves 5 feet (1.6 m) or more in height, this instrument measures only the low, long-period waves and ignores the much higher swell. The hole restricts the flow to a slow leak and its size can be "tuned" to the desired period. Such was the principle of Walter Munk's long-period wave (tsunami) recorder on Scripps pier, which first recorded surf beat. Electronics and software have revolutionized wave research.

WAVE RECORDERS

In the late 1940s, the Waves Project of the University of California installed twenty or thirty wave recorders off the California, Oregon, and Washington coasts, usually just outside the surf zone. Some operated for several years; others were knocked out in a few hours—the Pacific Ocean is a tough proving ground. But there were results, and the effort

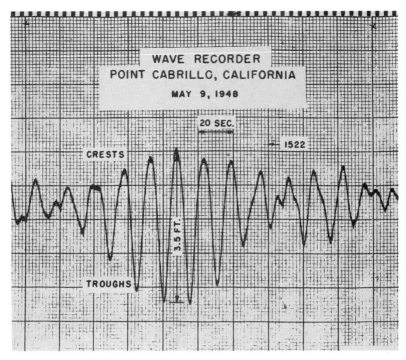

WAVE RECORDER
POINT CABRILLO, CALIFORNIA
MAY 9, 1948

20 SEC.

1522

CRESTS

3.5 FT.

TROUGHS

Early wave recordings were carefully selected from graph paper recordings, then examined with the eye and a ruler. All that has changed today. *US Army Corps of Engineers*

produced miles of chart paper covered with good records of waves, enough to form a sound basis for the first statistical summary of Pacific swell.

Early measuring/recording systems were generally similar. A differential pressure pickup on a steel tripod rested on the seafloor in 30 to 60 feet (9–18 m) of water; an armored submarine cable encasing three or four electrical conductors led ashore to a recorder in a shack on the beach. As the waves passed over, the sensor detected the changes in pressure at the bottom caused by the differences in the height of water above the instrument. The signals were electrically transmitted ashore, and on a moving chart a pen traced a red line representing the crests and troughs of the passing waves.

In most cases the transducer, a device that changes pressure into an electrical signal, was actuated by the motion of a diaphragm (a small metal or ceramic surface which flexes as the pressure changes). In the

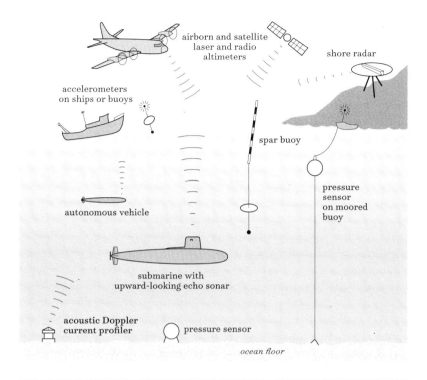

FIGURE 37: Deep-water and coastal wave-measuring methods.

tsunami recorder described earlier, the problem was to get rid of the effect of swell and record only long-period waves. But in these recorders, only the swell was important, and steps were taken to eliminate the effect of the long-period waves and tides.

An absolute or total pressure recorder would measure all waves (including swell and tides), superimposed one on another, plus the weight of the water above the instrument and the changing atmospheric pressure. A differential pressure sensor can be used to remove the effect of the tide.

The differential pressure instrument measures the rapidly changing pressure of the passing wave relative to the slowly changing pressure of the average sea level. Many subsurface wave-measuring instruments work in a similar manner. Software filters routinely support such data-processing needs. See figure 37 for a depiction of various wave measuring methods.

The bottom-mounted wave-pressure recorder has both advantages and disadvantages. Because it is installed usually in water 30 feet (10 m) or deeper, the higher-frequency waves are filtered out by the depth and do not confuse the record. That is, chop and small wind waves do not affect the pressure pickup on the bottom (because it is deeper than half their wave length). This is a disadvantage only to those who are interested in the small waves.

The bottom recorder requires no special installation offshore and can be placed almost anywhere on the bottom. On the other hand, it is harder to service and may be covered with sand in some seasons and thus rendered inoperable. A pressure record is not a precise reflection of the sea surface. It ignores the small waves, and the indicated heights of large waves must be corrected for changing depth and the speed of the passing wave (Bernoulli principle).

To overcome the misrepresentation of small waves, the wave-staff was created. Typically, the wave-staff detects the level of the water by measuring the resistance (or capacitance) of the water between two parallel wires or opposing electrodes. The output of the instrument is a direct record of the history of the height of the sea surface. Chop, wind waves, swell, and tides, each atop the others, are all recorded simultaneously in magnificent confusion. This method has now been modernized by using digital electronics, sensors, and data acquisition computers—but the confusion remains.

There is also interest in measuring the waves in the ocean far from the shore where the depth of water is so great that a bottom recorder would detect only the very long waves. It is possible to make use of this filtering effect of depth selectively to record tides and tsunamis. Another method was to mount a shallow-water-type pressure recorder on top of a submerged buoy held about 100 feet (30 m) below the surface by a taut wire anchored to the bottom.

A more commonly used method of measuring waves in deep water is to mount accelerometers in a floating buoy (moored or free-drifting) and directly record the acceleration of the buoy caused by the passing waves. The ship can take the place of the buoy, but because a ship is large compared to the waves, a combination of accelerometers and pressure pickups at the bow and stern is required. The instruments

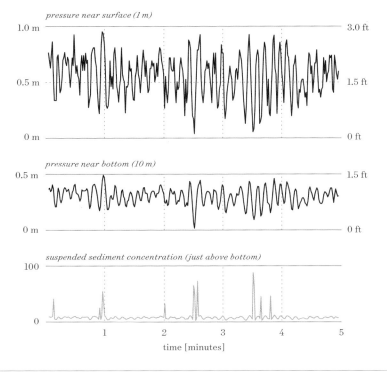

pressure near surface (1 m)

pressure near bottom (10 m)

suspended sediment concentration (just above bottom)

time [minutes]

FIGURE 38: Pressure measurements of passing waves (in black) are irregular near the surface; near the bottom they diminish in size and smooth out. At the bottom, the larger wave groups stir up the sediment (in blue) with their horizontal movements.

are mounted in the hull about 10 feet (3 m) below the normal water level. The pressure sensor measures the height of passing waves above this point, and the accelerometer measures the height of the pressure pickup relative to average sea level. The signals are fed into a computer that sorts the data and records the major waves (see figure 38).

Since the modern study of waves began in 1946, a cornucopia of inventions has enhanced wave measurements. In the late 1940s, the o-ring (a spin-off of the aviation and fluid-pumping industries) increased the designers' confidence in making watertight pressure vessels. The commercial advent of the transistor (1950s) began an onslaught of electronic devices that has almost removed humans from the surf zone. The microprocessor (1970s), GPS (1980s), and digital signal processors (1990s), all now connected by the internet, allow for rapid data collection, processing, and warning. The clipboard and graph paper disappeared; the stadia rod, measuring tape, and transit all became obsolete in the surf zone. Today, GPS-equipped drifters can meander with the long-shore currents as dye-stained parcels of water are tracked by cameras and pattern-recognition algorithms. We have mountains of data, yet many ocean researchers never touch the ocean. But there is always something to personally observe at the ocean's edge—you just have to leave the promontory of the keyboard—something will always be there for you to observe. Go and see.

What follows is an overview of the instruments that are readily available in the twenty-first century. Light waves, radio waves, sound waves, and GPS are all used to measure the watery world.

The Use of Light (Optical)

There are several techniques for studying and measuring waves and bathymetry from a known altitude. One is the laser altimeter, which uses a laser (coherent light) beam pointed downward. The laser pulses reflect off the surface of the waves and from the sea bottom, and the timing of reflections can provide precise information about wave heights and water depths. The airborne use of lasers, sometimes called Airborne Laser Bathymetry, is an offshoot of 1960s submarine-detection methods. By the 1980s, similar methods were being used in the surveying of the coastal regions and recently with small, remotely piloted aerial drones. The combination of lasers, optics, and electronics has evolved into Lidar (Light Detection and Ranging). Lidar can accurately and rapidly measure distant objects by using the time-of-flight of pulses of light (hundreds to thousands per second) resulting in an image of the coastline and ocean (Google Earth utilizes Lidar data).

Another optical instrument, the optical backscatter sensor (OBS), revolutionized suspended sediment and water turbidity (clarity) measurements. John Downing, while still a PhD student at the University of Washington, developed the instrument in the late 1970s. It exploits the optical backscattering (reflection) of infrared light off of the suspended sediment particles in the water column. When the reflected light reaches a photodiode, the OBS outputs an electrical signal proportional

September 16, 2003   September 23, 2003

Frisco, NC

9m

-1m

high dune

low dune

buildings   beach

Frisco, NC

dune breach

two buildings lost

LIDAR Lidar has revolutionized coastal measurements. Hatteras Island, North Carolina, was remotely imaged before and after Hurricane Isabel (September 2003). A photograph is on the right for comparison. The breach took two months and millions of dollars to repair. *Asbury Sallenger and Amar Nayegandhi, USGS Center for Coastal and Watershed Studies*

to the suspended sediment load, once again making it easier for ocean researchers to collect data while remaining on dry land.

A more complex and expensive method is Laser In-Situ Scattering and Transmissometery, which measures both particle size and concentration of suspended sediments. We will leave the more detailed explanation of how it functions to other publications.

The Use of Radio

Another innovation is the use of Radio Detection and Ranging (referred to more commonly as radar) to measure surface waves and currents. A series of pulsed radio emissions is backscattered by waves (from capillary waves to waves of up to 330 feet, or 100 m, in wave length). The wave lengths measured are about half that of the radio-wave wave length used, and the returning echoes from each pulse are received and processed electronically and mathematically to make an image of the

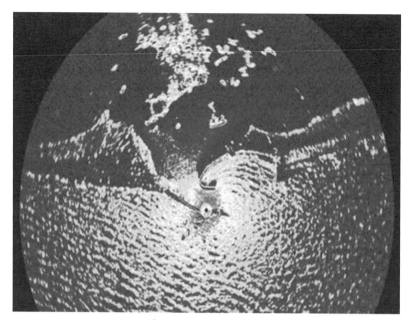

MARINE RADAR Marine radar exploits 'sea-clutter' to produce images of wind, waves, currents, and bathymetry much like weather radar informs us of weather fronts. This marine radar image was made at the port entrance to Höfn, Iceland, in January 2019. Shear in the water column can be calculated from radar data. *Jochen Horstmann, Helmholtz-Zentrum Geesthacht*

sea surface. This technique is used to obtain wave length, wave direction, and surface currents. The techniques that use radar are frequently referred to as HF radar (or high frequency radar, although not all operate at high frequencies). Radar systems cost tens of thousands to hundreds of thousands of US dollars to purchase.

Further aloft from space, very expensive satellites equipped with altimeters and synthetic-aperture radar techniques use radio pulses to measure sea level, significant wave heights, and wind speeds, and to calculate wave energy with surprising accuracy. Amazingly, sea level can be determined to less than one-tenth of an inch (3 mm) for processed, averaged data and less than that with careful data analysis.

THE USE OF SOUND

The advent of the acoustic Doppler current profiler (ADCP) in the early 1980s was a revolution.

The acoustic Doppler current profiler (ADCP) sends out pulses
of sound from each transducer (here in red), then listens for
their Doppler-shifted reflections. *Kim McCoy*

Acoustics are now routinely used to measure and record water and
sediment movements. The ADCP works similarly to the lidar and radar
techniques described previously, but the ADCP uses acoustic waves (instead of light or radio waves) to sense ocean currents and waves. The
ADCP uses a piezoelectric ceramic sensor, similar to a buzzer, which
can be made to vibrate by electricity. Interestingly, the process is re-
versible such that when an abrupt mechanical force (such as a sound
wave) strikes a piezoelectric material, it produces electricity! A microphone exploits this phenomenon. The ADCP sends out a high-frequency
sound (tones). The tiny particles in the water column reflect some of
the sound back to the ADCP. If the ADCP is positioned to look up, surface reflections will be received. A whole bunch of complex electronics,
DSPs, mathematics, clever software, and microprocessors inside a watertight enclosure are "made simple" for the user for less than $10,000
USD. The various types of ADCPs (also called acoustic Doppler veloci-
meters) can measure the movement of a parcel of water between the

size of a sugar cube to the size of a car garage in volume. Deep-water and long-period waves can be measured from the depth of more than 3,300 feet (1,000 m) for over a year; producing an immense amount of data. The acoustic backscatter sensor (ABS) uses the intensity of reflected sound to measure the sediments moved by waves and currents (see figure 38, page 197).

Even DUKWs have been replaced. Today, personal watercraft equipped with a GPS receiver and echo sounder can rapidly survey nearshore sandbar migration (see photo on page 231). It works like this: The watercraft's position is accurately updated several times per second (in three dimensions) by using GPS. At the same time, the echo sounder measures the height above the sandy bottom. Combining the depth of the water below the watercraft with the GPS location produces a profile of the sand. When everything works, a fast and efficient sandbar profile can be produced sometimes even just before and after a storm.

Collecting beach data into databases has transformed most researchers from being wet and cold into perfectly comfortable computer programmers and data processors who spend most of their time at their desk. Going to sea is a decreasing requirement and is increasingly expensive. Because of these developments in instrumentation, we now have reams of data extending from infragravity waves, ocean swell, breaking waves, sediment bed loads, and inner surf zone set up into the swash, which provide an almost complete energy-density spectrum for ocean waves. This is just an overview of today's instruments and methods. Harald Sverdrup, John Isaacs, Roger Revelle, Walter Munk, and all seagoing pioneers in oceanography never dreamed it would be so easy. Although purchasing an instrument is relatively simple, careful planning, deployment, and retrieval is still necessary to collect meaningful data.

WAVE-FORCE MEASUREMENT

Another class of instrument makes it possible to measure the force exerted by waves on pilings, piers, and shoreline structures. Some instruments have been used to observe storm wave forces and obtain data that can be applied in the design of offshore structures, including large oil-drilling and production platforms. Others are intended to determine the shock caused by the impact of a breaking wave on a breakwater or another very-shallow-water structure.

The measurement of wave forces on pilings is complicated by the continual reversal of direction of the water as the crest moves in one direction and the trough moves in the other. The water velocities in the various parts of the wave vary with time and with depth. Moreover, because the force is caused by the rush of water past the pile (the drag), the answer is sensitive to the square of the velocity as well as to the shape and size of the piling. There were so many unknown factors that, before the availability of computers, it was best to determine the answers directly by experiment: first in the model tank, then at sea. In the early tests, specially instrumented pile sections were exposed to ocean waves under various sets of conditions. Today, computers can model wave forces on structures faster and less expensively than sea testing. We have models for many purposes, including deep-water waves, refracted waves, and even shoaling waves for surfers. Instrumented-at-sea testing is used to validate the performance of computer-designed structures. The forces imposed on pilings by swell are relatively modest, but as the wave begins to break and the water-particle velocity increases, waves of about the same period and height cause substantially greater forces. One moral of this story: when possible, place your structure in water too deep to cause waves to break.

Early wave-direction recorders did not work very well, the reason apparently being that waves from so many directions are always present and the direction sensor is as confused as any human observer would be.

Many structures are now constructed, bristling with embedded sensors, including strain gauges, to measure compression, tension, and torque, rather than directly measure the wave forces. Today's methods utilize three or more sensors and precisely correlate the time of passage of the wave crests to obtain wave direction.

In turbulence research (the study of unsteady fluid movement), direct measurements are still made by using shear probes. A shear probe is made with a piezoelectric beam, which, when bent, outputs

Predicting the shock forces of breaking storm waves is complex
yet essential to the survival of offshore and coastal structures.
Borgholm Dolphin platform, North Sea. *Richard Child courtesy
of Creative Commons*

an electrical signal. It is normally shaped like the tip of a bullet. Everywhere water flows, turbulence occurs: over a ship's propeller, within a
breaking wave, over a barnacle on a rock, below a shoaling ocean swell
and with the rise and fall of the tides as tidal friction. Nature is always
in the process of spreading out its energy. Turbulence, although persistent and observable in many small-scale oceanic processes, is usually
described with sophisticated mathematics.

Most oil platforms are deployed in sufficiently deep water to avoid
shoaling breakers, but when great hurricanes sweep across the Gulf of
Mexico, the North Sea, and other stormy regions, the structures must
withstand breaking waves at sea. Actual waves have random shapes
and speeds; rarely are they uniform along a crest. Offshore-design engineers are compelled to understand waves, reduce construction costs,
and ensure platform survival.

In the United States, the Army Corps of Engineers is responsible for the maintenance of harbors and coastal structures. It has conducted experiments to determine how great shock pressures can be and how new structures might be designed to resist them. Anyone who has stood on a rocky coast in a storm has felt the ground shake under the waves and seen the water hurled high into the air and will have some appreciation of the problem. The pressure required to project water into the air is about one-half pound per square inch (30 millibars) for each foot (30 cm) of height. Thus, water going 40 feet into the air requires 20 psi (10 meters high requires 1 bar of pressure) or an impact load of nearly 1.5 tons on each square foot of rock.

The shock-pressure gauge consists of a stack of piezoelectric sensors set in a strong casement. When subjected to pressure, this gauge produces a small charge of electricity that can be recorded by computers. In wave-channel experiments, breaking waves only 6 inches (15 cm) high have produced pressures as high as 18 psi—but lasting only 0.001 of a second. You have to be fast to measure breaking waves.

The harshly energetic surf zone has not changed since researchers surfed the waves in DUKWs in the 1940s. Despite our great sums of data, the ocean's waves can still destroy structures, erode coastlines, or even kill you.

Making Waves

There are many kinds of wave makers and many uses for model waves in controlled conditions. Experimental facilities range in size from tabletop ripple tanks to maneuvering basins much larger than a football field. But why make experimental waves? There are many reasons, and answers are needed to the questions: How are certain waves created? What shape are the orbits under various conditions? How are they propagated? What are the conditions under which waves of different sizes refract, reflect, and diffract?

The shoreline engineer must know in advance of construction the effects of various kinds and sizes of waves on beaches, breakwaters, groins, jetties, seawalls, and similar structures. How effective will these structures be in protecting a harbor or stopping the drift of sand along

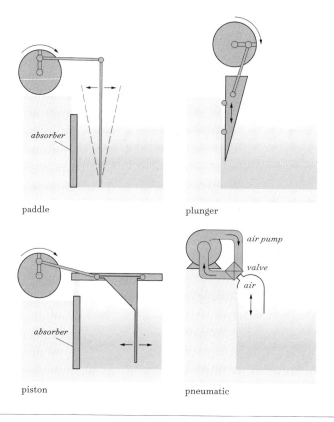

absorber

paddle

plunger

air pump

valve

air

absorber

piston

pneumatic

FIGURE 39: Four types of model tank wave makers.

the coast? How high should a new dam be to prevent storm waves from going over the top? What size rocks should be used in the breakwater?

The naval architect wants to tow ship models in wave conditions simulating a true seaway and determine the magnitude of the stresses on each part of the hull. The designer can discover also how seaworthy a design will be and can estimate the speed that some future ship will be able to make into the teeth of a gale that will blow a dozen years hence.

The several types of wave makers in general use are shown in figure 39. No particular ingenuity is required to produce waves in a tank; in fact, it would be a much more remarkable feat to do anything to the water without making waves. Each type has advantages for special applications. The paddle, the plunger, and the piston are all connected by a rigid arm to an eccentrically located pin on a turning wheel and thus directly produce mathematically satisfying sine waves. One can see by inspection of the diagrams how these work. It is evident that reducing the speed of the driving wheel lengthens the wave period, and increasing the radius of the pin connection increases wave height. These are commonly used in long, narrow wave channels to produce a large variety of wave sizes.

Pneumatic wave makers are mounted side by side along two walls of large, square tanks. They create waves by changing the air pressure beneath a hood so that the water surface there rises and falls. As the water surface inside the hood is depressed the pressure is transmitted, according to Pascal's law, through the water to the water immediately on the other side of the partition where the surface is raised. This disturbance will then travel the length of the tank. The speed of the blower motor controls the amount of air pressure and thus the amplitude of the waves; the speed of the valve-operating motor controls duration of pressure and thus the wave length.

The formation of a wind wave starts when the wind imparts some of its energy to the water. This process is very important to the understanding of how capillary waves grow into surface gravity waves. It is difficult to make measurements at sea and even harder during a hurricane—laboratory wave-making tools help for understanding the theory before making the big, expensive, and sometimes catastrophic plunge into the sea.

In most cases it is simpler to make waves than to find waves. Many of the large wave-making facilities have been dismantled in favor of computer modeling. However, some old ones do remain, and new ones have been built. There are two wave channels at the Scripps Institution of Oceanography Hydraulics Laboratory: one wind wave channel and one glass-walled wave channel. Both can generate waves of controlled periods and heights. The wind wave channel can create waves by varying the wind speed over its 145 feet (44 m) in length, 7 feet (2 m) in width, and 7 feet (2 m) in depth. The University of Hamburg in Germany has a similar wind wave channel. The University of Hannover's Coastal Research Centre has a 1,000-foot (330-m) long wave flume—the world's longest.

The new Scripps Ocean Atmosphere Research Simulator is scheduled for completion in 2020. This fully enclosed wind wave channel will be able to simulate the interactions of the atmosphere and ocean. It will be unique in its ability to vary marine spray (aerosols) from the surf, atmospheric gases, and pollutants that influence our changing climate.

Many years ago, at the University of Kenya, there was no money for a mechanical wave generator. Instead, one of the students rocked a piece of plywood, whose lower edge rested on the tank bottom, back and forth between two chalk lines on the tank rim, timing the action with the second hand of a watch. As far as I could see, it worked about as well as most other generators and, of course, it was instantly variable at the professor's request.

It is also easy to generate capillary waves and compare their mechanics with those of gravity waves. A very small drop of water landing on a still-water surface will generate surface wrinkles (capillary waves) that radiate outward. As the size of the drop is increased, the capillary waves will be seen to be followed by tiny gravity waves that plainly have a longer wave length. An eyedropper and a bathtub are all the equipment needed.

At Duck, North Carolina, a research pier extends 1,800 feet (550 m) seaward to 20 feet (6 m) of water. It is instrumented with various wave gauges and a radar that obtains wave direction—the principal objective being to discover the changes in the character of waves as they cross the shelf. The Offshore Technology Research Center, in College Station, Texas, has a basin that is 150 feet (46 m) long and 100 feet (30 m) wide. This clever wave maker can simulate ocean waves from several directions, gusty winds, and sheared current profiles all at once. This facility has helped in the design of many deep-water tension leg platforms, spar buoys, and remotely operated vehicles for the oil industry. Even the National Aeronautics and Space Administration has used the facility to help explore its Assured Crew Return Vehicle concept for the International Space Station.

HR Wallingford in Great Britain, the Shanghai Ship and Shipping Research Institute in China, and the National Research Council Canada's Multidirectional Wave Basin are a few more of the many excellent wave-making facilities around the world. There are a great many more wave channels, test basins, hydrodynamics laboratories, and experimental wave facilities, but covering them in any more detail would be beyond the scope of this book.

Transforming Waves into Knowledge

The use of mechanical wave energy took an interesting new path with the development of the Wave Glider by the company Liquid Robotics (founded in 2007). The Wave Glider is about the size of a large surfboard, and cleverly converts the waves' orbital motions into forward propulsion by using a combination of mechanical engineering, bungee cords, hydrodynamics, and electronics. The Wave Glider has become a powerful tool for ocean researchers. It bristles with an array of sensors, hydrophones, and telecommunications as it journeys across the world's oceans. The autonomous vehicle has navigated from San Francisco, California, to Queensland, Australia, a distance of almost 8,000 nautical miles (14,000 km) powered only by the waves!

Another wave-powered vehicle is the Wirewalker, which exploits the vertical motion of the waves as it pulls itself downward along a cable. The cable is held taught by a weight as the Wirewalker (with a suite of sensors) travels into deeper water. At the bottom, a clever mechanism releases the vehicle from the power of the waves and allows a graceful buoyant trip to the surface. The process, which uses no batteries or motors, can profile vertically thousands of times.

As described earlier, internal waves exist where there is strong change in density frequently caused by a change in temperature. An autonomous underwater vehicle called the Slocum Thermal Glider cleverly exploits the ocean's temperature differences for propulsion. Although much more complicated, it uses the difference between the upper-ocean water temperature and the cooler deep water to cause an onboard change in volume to vary its buoyancy. The thermal glider has flown thousands of miles—propelled by exploiting ocean thermal energy. These are smart and renewable methods that change wind-, wave-, and temperature-based energy into scientific knowledge.

9   The Surf

Waves have many stages in their lives: they are born as ripples, grow into whitecaps, chop, wind waves, and finally, into fully developed storm seas. As these seas pass from under the winds that formed them, they diminish in height and steepness into low sine-wave-shaped swell. As swell, waves may traverse great stretches of open ocean without much loss of energy. Eventually they reach the shoaling waters of a continental shelf. On the shelf, the wave fronts are bent until they are almost parallel to the shoreline. All this seems to be merely preparation for the final and most exciting step.

The irregular waves of deep water are organized by the effect of the bottom into long, regular lines of crests moving in the same direction at similar velocities. The romanticist thinks of the forces of the sea being marshaled for an exuberant death against an ancient enemy. The depth continues to decrease until finally in very shallow water when it becomes impossible for the oscillating water particles to complete their orbits and the battle begins. When the orbits cease, the wave breaks. The crest tumbles forward, falling into the trough ahead as a mass of

PREVIOUS SPREAD: A wave seen at the instant it transforms from shoaling into breaking. From this point to the top of the beach face is the surf zone. Mentawai Islands, Sumatra, Indonesia. *Ted Grambeau* foaming white water. The momentum carries the broken water onward until the wave's last remaining energy is expended in a gentle swash that rushes up the sandy beach face and sinks from sight. The inertia of the wave is gone! This zone where waves give up their energy and where systematic water motions give way to violent turbulence is the surf zone. It is the most exciting part of the ocean.

Breaking Waves

As the swell from the deep sea moves into very shallow water, it is traveling at a speed of 15 to 20 mph (25–30 km/h), and the changes in its character over the final few dozen yards to shore come very rapidly. In the approach to shore, the drag of the bottom causes the wave velocity to decrease. The decrease causes the phenomenon of refraction, which was described starting on page 103, and one of its effects is to shorten the wave length. As length decreases, wave steepness increases, tending to make the waves less stable. Moreover, as a wave crest moves into water whose depth is about twice the wave height, another effect is observed that further increases wave steepness. The crest peaks up. That is, the rounded crest that is identified with swell is transformed into a higher, more pointed mass of water with steeper flanks. As the depth of water continues to decrease, the circular orbits are squeezed into a tilted ellipse and the orbital velocity at the crest increases with the increasing wave height.

This sequence of changes in wave length and steepness is the prelude to breaking. Finally, at a depth of water roughly equal to 1.3 times the wave height, the wave becomes unstable. This happens when not enough water is available in the shallow water ahead to fill in the crest and complete a symmetrical waveform. There is too much energy and too little water. The top of the onrushing crest becomes unsupported and it collapses, falling in incomplete orbits. The wave has broken; the result is surf.

Having broken into a mass of turbulent, tumbling foam, carried landward by its own momentum, the ex-wave will, if the water deepens again as it does after passing over a bar, reorganize itself into a new wave with systematic orbital motion. This reorganization is the result of dumping the mass of water from the wave crest into the relatively quiet water inside the breaker zone; the impulse generates a new wave. The new wave is smaller than the original one, the difference in heights representing the energy lost in breaking. The new wave, being smaller, proceeds into water equal to 1.3 times its new height; then it, too, breaks (see figure 40).

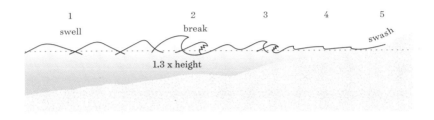

1   2   3   4   5
swell   break   swash

1.3 x height

FIGURE 40: The figure shows the anatomy of a breaking wave: (1) swell peaks up on entering very shallow water; (2) at a depth equal to 1.3 times the wave height, it breaks; (3) wave reforms and breaks again on an underwater bar; and (4) water moves beachward as a wave of translation.

Again, a mass of water is produced—white with bubbles of entrained air—but the water is likely to be too shallow for a new oscillatory wave to form. Now the front of the water becomes a step-shaped wave of translation—a different sort of wave in which the water actually moves forward with the waveform rather than merely oscillating as the waveform passes. Finally, at the beach face, the momentum of the water carries it into an uprush; the water slides and sprawls in a thin swash up and across the face of the beach. As it reaches its uppermost limit the wave dies; all the energy so carefully gleaned from the winds of the distant storm and hoarded for a thousand miles of ocean crossing is gone, expended in a few wild moments. Because the energy is released so rapidly, the energy density in the surf is actually much higher than in the storm, which originally created the waves.

The surf changes from moment to moment, day to day, and beach to beach. The waves are influenced by the bottom and the bottom is changed by the waves. And because the waves arriving at a beach are highly variable in height, period, and direction, each wave creates a slightly different bottom configuration for the ones that come after it. The water level changes with the tide, and the waves change as the storms at sea develop, shift position, and die out again. The result is that the sand bottom is forever being rearranged. Even in glass-sided wave channels where an endless number of waves, each exactly the same, can be produced, equilibrium is never reached; the sand continues to change as long as the wave machine is running.

Thus, the waves change the sand on the bottom at the same time the sand is changing the waves. First, consider the effect of the bottom on the waves as they break. It may make them *plunge, spill, collapse,* or *surge*. Plunging breakers are the most impressive. Their principal characteristic is very rapid release of energy from a wave moving at high velocity. There is a sudden deficiency in water ahead of the wave, which causes high-velocity currents in the trough as the water rushes seaward to fill the cavity beneath the oncoming crest. When there is not enough water to complete the waveform, the water in the crest, attempting to complete its orbit, is hurled ahead of its steep forward side and lands in the trough. This curling mass of falling water will often entrap air and then, as the upper part of the wave collapses, the air is compressed. When the compressed air finally bursts through the watery cap, a geyser of water is hurled into the air—sometimes over 50 feet (15 m).

If there is a strong offshore breeze, the thin crest of the wave will be blown off as it plunges forward, leaving a veil of rising spray behind to mark the path it has followed. This delicate tracery of spray has been likened by poets to the "white manes of plunging horses." Anyone who has observed such breakers, backlit by a low Sun, will understand the comparison and agree that this circumstance is worthy of poetic description.

To understand the reasons why breakers plunge, it calls for a somewhat more scientific approach. The wave must retain most of its energy right up to the moment of breaking. That is, there should be nothing, such as a rough bottom, a strong wind, or substantial currents, to make the wave prematurely unstable. Any of these conditions will degrade a wave's energy by slowing it down and warping its orbits to make it break gradually rather than abruptly. Thus, when a large, clearly defined

PLUNGING WAVE A plunging breaker peaks up and curls over as it encounters a rapid change in depth, such as over a bar or reef. Pipeline, Hawaiʻi. *Brian Bielmann*

swell passes over a steep, smooth underwater slope of the proper depth on a calm day, a perfect plunging breaker will result. If, however, the bottom is gently sloping and studded with rocky irregularities, or if the approaching waves appear confused, a spilling breaker is more likely to be produced.

A spilling wave breaks slowly and without the violent release of energy needed to fling the crest forward into the trough ahead. Its crest merely tumbles down a more gently sloping forward side, sometimes over a considerable distance and lasting for several minutes. Therefore, spilling breakers are much favored by novice surfers, who ride on the face of the wave, their boards doing much the same thing the tumbling white water is doing.

The famous surfing area at Waikiki, where surfers in outrigger canoes and on boards are frequently photographed against Diamond Head, is a fine place to observe perfect spilling breakers. There, a very shallow, gently sloping coral reef extends for 1 mile (1.6 km) outward

SPILLING WAVE A spilling breaker moves forward, taking longer to break across a greater distance as it passes over a gentler change in depth. *Tamara K./Deposit Photos*

from the beach. The tide range is very small so that almost all swell coming in from the Pacific is converted into low, spilling breakers. The surfer can paddle out as far as desired and be assured of a ride back.

Many surf zones have larger tidal ranges and are underlain by shifting sand; instead of remaining the same as Waikiki does, their underwater topography is constantly changing. The result is that in most surf zones one can observe some combination of plunging and spilling breakers, and forms that are intermediate between the two. That is, the breaker plunges, but without sufficient momentum to hurl the water beyond the sloping forward side into the trough ahead. The falling curtain of water lands partway down the wave front, and the breaker has an intermediate section.

Some beaches have such a steeply sloping approach that a swell nears the shore without being slowed or changed until the last possible moment. Then it will abruptly rise up and break directly on the beach face or reef with astonishing violence. These are collapsing breakers.

In special cases, when the wave height to wave length ratio (steepness) is small and the beach face is steep, long-period waves will surge up the steep beach face without breaking.

In other areas the beach approach may shoal so gradually that the surf zone may be as much as a mile wide. On the beaches of Oregon and Washington that have underwater slopes of about 1:100 (1 foot vertical to 100 feet horizontal) it is not unusual to have three lines of breakers when the great winter storm waves arrive. The outer line of breakers may be plunging, 30 feet (9 m) high, with sufficient violence to shake earthquake-measuring instruments several miles inland. Having broken, they will reform and break twice more with decreasing violence as they cross a half mile of irregular shallows to reach the shore.

The movement of broken waves in shallow water creates another type of wave: the wave of translation. This wave was first studied by John Scott Russell in the 1840s. The name "solitary wave" is now used by mathematicians to describe this phenomenon.

Two striking characteristics clearly differentiate the wave of translation from the ordinary oscillatory waves that we have been considering. First, the entire form of this wave is above the undisturbed water level; that is, it consists of a crest without an accompanying trough. Second, there is an actual translation of the water particles as the waveform passes. An object floating in the water would be carried a definite distance forward by a wave of translation and come to rest, without exhibiting the corresponding backward motion observed in the wave of oscillation.

Russell found that the wave of translation is produced by the sudden addition of a mass of water to a still-water surface. When the oscillatory wave breaks, the water in the broken crest falls onto the water surface in advance of the oncoming wave, producing a wave of translation or a step-shaped foamline, which continues shoreward. Hence, although this type of wave is nonexistent in the open, unconfined sea, it becomes important in the shallow waters inside the breakers, where most oscillatory waves eventually are transformed into waves of translation.

These waves travel at the velocity $C = \sqrt{gd}$ (g is the force of gravity and d is the depth of water) with the unusual variation that, because the wave height is large compared to the depth of water, the two are added

COLLAPSING WAVE A collapsing breaker suddenly walls up over an abrupt change in depth. Such waves are beloved by expert surfers. Most of its energy, accumulated over great distances, is immediately lost in turbulence. Teahupo'o, Tahiti. *Ted Grambeau*

to give d. The velocity, therefore, is related to the water depth, and when there are several waves of translation moving shoreward at the same time, the later ones move faster and tend to overtake the ones ahead because they are traveling in deeper water on top of their predecessors. Solitary waves are observed from space and are frequently observed at the Strait of Gibraltar at the entrance to the Mediterranean Sea.

Surf Beat

One of the wave characteristics most evident to an observer standing on the shore is the variability in height of the breakers. A series of a dozen or so low waves will approach and break. Then there will be a group of several high waves—usually three or four in what is called a wave set—then another relatively quiescent period.

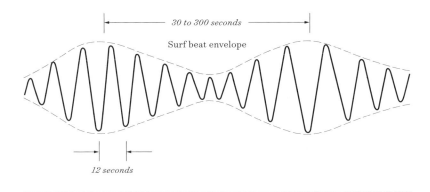

FIGURE 41: Two groups of waves, each about twelve seconds in period, combine to reinforce and cancel each other, shown causing a surf beat with a period of two minutes.

Sometimes this variability is caused by the arrival of two sets of swell (from two storms) of nearly the same period at the same time. When the crests of the two wave trains almost coincide, they reinforce each other and produce waves higher than those of either set. When the waves are almost completely out of phase and the crests in one train coincide with troughs in the other, the resulting waves are small. As the phase relationship changes, a pattern emerges like that shown in figure 41. The envelope of these wave trains (shown dotted) has a waveform and a period or "beat frequency"—usually 30 to 300 seconds—but it is not a true long-period wave.

The effect of groups of breakers of alternating height is to raise and lower the average water level in the surf zone. The rise is somewhat exaggerated because the volume of water transported into the surf zone is proportional to the square of the breaker height.

Anyone wading in shallow water notices that a rapid succession of high breakers temporarily raises the water level. Sea-level variations of as much as 16 feet (5 m) have been observed at Twin Rocks, Oregon, caused by surf beat in very heavy weather. The waves of translation resulting from the large breakers transport a considerable volume of water shoreward on the surface faster than it can escape seaward again along the bottom.

The consequence of this process is that the shoreline tends to act as a source of new waves that return about 1 percent of the incoming wave energy seaward as true long-period waves. These newly formed waves resulting from surf beat move seaward and along the coast, and they can cause the surging (seiches) in harbors as discussed in Chapter Six.

Undertow and Rip Currents

One of the most ubiquitous concerns at the seashore is that of *undertow* or *rip current*. The very words frighten many would-be surf swimmers, and some beaches have signs that say, "Dangerous Undertow, Swim at Your Own Risk." In most cases, warnings pertain to a rip current and not undertow, although "undertow" is still used.

There are currents flowing in the surf zone, and there are other water motions that may cause trouble for swimmers, but they are frequently and improperly interchanged. Consider them one at a time. Orbital currents in the waves perform circles equal to the height of the waves with the period of the waves. A swimmer in waves performs these circles as the water does: Half the time, these move the swimmer down to seaward; the other half, up to landward. After each wave passes, the swimmer is back around the same spot as at the beginning of the wave. If the swimmer gets in the trough of a breaker, the swimmer will indeed get sucked under it, and as it breaks, the swimmer will be upended and propelled landward, possibly to be cast up on the sand.

The foamlines of broken water (a type of wave of translation) do transport water landward that must somehow return seaward as a current. If the beach has a reasonably even slope inside the bar, there may be a return current on the bottom. But in order for the wave of translation to endure, the water must be quite shallow—perhaps 2 or 3 feet (60–90 cm). So, in most circumstances, swimmers could stand on the bottom and, even if knocked down by the water moving landward, they would certainly not be carried out to sea along the bottom by the relatively small undertow return current.

On very steep beaches where large waves break directly on the beach, the uprush and backrush may be violent surges of water as much as 2 feet (60 cm) deep. These are quite capable of knocking a person down and rolling them back down the beach into the path of the next breaker, where they could be mauled and tossed about by several waves

RIP CURRENT A rip current moves seaward in a channel between two bars. Rips will rapidly transport sediments, surfers, and swimmers offshore. It will also influence the oncoming waves. *David Clark/Woods Hole Oceanographic Institution*

before regaining footing. In such circumstances usually the best escape is to swim seaward and reach the calm water outside the breakers—it is not far away. When there, the right wave can be selected and ridden in, sliding far up the beach on the uprush, digging into the sand at the high point, and holding there until the backrush draws the water away; then scrambling to high ground before the next uprush. This maneuver can be real sport, though it is a bit dangerous. However, steep beaches with high surf are rare; certainly, there are none in the usual resort areas. But this minor hazard does not fit the popular description of "undertow."

There is one form of current in the surf zone that can be dangerous to an inexperienced swimmer. This is the rip current, first described by Professor F. P. Shepard of the Scripps Institution of Oceanography. Also known as the "rip," it is of great importance in beach processes,

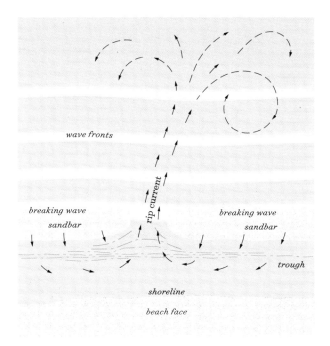

wave fronts

breaking wave

sandbar

rip current

breaking wave

sandbar

trough

shoreline

beach face

FIGURE 42: Rip currents are created when waves break on a shallow bar, making waves of translation that raise the water level shoreward of the bar. The excess water then flows seaward as a swift and narrow rip current in a channel of its own making, sometimes ending in a vortex outside the surf.

for it is responsible for some of the strange forms that the underwater parts of a beach may take.

Rip currents are created when waves break on and flow over a shallow bar, making waves of translation (foamlines), as shown in the photo and in figure 42. These waves raise the water level ever so slightly, just above sea level, shoreward of the bar. The continual wave-driven supply of excess water over the bar seeks to return to the open sea. It first flows parallel to shore then at the lowest part of the bar flows outward. Now, the swift and narrow (rip) current can push seaward in a channel of its own making, sometimes ending its life in deeper water by shearing off into a swirling vortex outside the surf. These pinched-off, slowly rotating forms are called shear waves, which can interact with coastal edge waves (as described in Chapter Four). Such swirling may be enough to get your head spinning yet they are found in the ocean, the atmosphere, and throughout the universe.

Because the depth of water is greater in the rip channel than over the bar on either side, waves rarely break in the channel. Moreover, a current flowing against the waves has the effect of increasing wave steepness. The crests become prematurely unstable and a small spilling breaker may result, or, more likely, a large number of short, steep waves will develop that look something like wind chop. The result is that rip currents usually can be seen from the beach, especially if one can observe the surf zone from a vantage point.

The fact that large waves are less likely to break in the rip may actually encourage swimmers to choose the zone of high currents. Anyone who experiences being carried outward should not try swimming shoreward against the strong current but should swim to one side or the other—usually a short distance—and get out of the rip current to where the effect of the breakers will carry the swimmer shoreward again.

The following thoughts may be helpful to would-be surf swimmers, especially those who do not consider themselves to be very strong swimmers. The surf can be a dangerous place, for breaking waves produce sudden violent forces and swift currents. Therefore, before you plunge in and eagerly try to swim out through the breakers, you would be well advised to take a few minutes to look the situation over. Breakers vary considerably in size, but the high ones often come in groups several minutes apart. So, stand at a proper vantage point and just watch what's happening for, say, five minutes.

Watch the waves break—that's where the bars are. And don't forget that these bars are shallow; you may be able to stand on the bottom in water only waist-deep well out beyond places that are over your head. Generally, the lighter-colored foaming water is shallower than the darker water, and it may mark a place where you can rest for a while.

If one is studying waves from the shore, or if you must decide whether to venture out into the surf in a boat, it is useful to be able to accurately judge the height of the breakers in advance. This is easily done, even if the line of breakers is well offshore. Simply stand on the beach face at such a level that the top of the breaker is exactly in line between your eye and the horizon. Then, as shown in figure 43, the vertical distance between eye and backrush curl (which is about at the same level as the average sea surface) is equal to the height of the

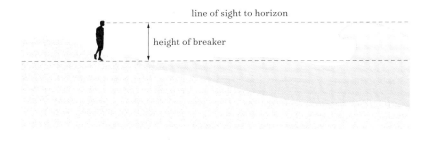

line of sight to horizon

height of breaker

FIGURE 43: When the observer's eye is aligned with the top of the breaker and the horizon, the vertical distance between the eye and the backrush is equal to the height of the breaker.

breaker. You may be surprised at how high the larger breakers really are, but it is a lot better to be surprised on the beach than in their midst.

So check the height of the breakers, look for rip currents, and enjoy the show.

SURVEYING PIONEERS IN THE SURF

In 1945 I had never seen the ocean, but I joined the World War II Waves Project of the University of California at Berkeley. The project had been set up to develop scientific means of determining the characteristics of beaches and of the waves that would make it difficult for landing craft to approach enemy-held beaches. Studying the surf was our business. Later, when the war was over, the project continued, its objectives becoming more broadly scientific.

M. P. O'Brien, dean of engineering and a member of the Beach Erosion Board, directed the work and, almost on the day I joined the project, he ordered our field party north to study the waves and beaches of Northern California, Oregon, and Washington. His theory was "If you can work there, you can work anywhere." Subsequent experience certainly proved him correct.

So, we set out for the northern beaches, timing our arrival to coincide with that of the great waves from the winter storms. John Isaacs, an old hand along the coast, was party leader; I was his engineer assistant. Field-party equipment included a collection of aerial cameras, walkie-talkie radios, a Catalina PBY flying boat, and two DUKWs.

(DUKWs are formidable on the highway, and even the log trucks would give them a fair share of the road, but in the surf they seem no more than a chip of wood. In the water they move by means of a screw propeller, and fortunately they turned out to be probably the world's best surf craft, or else I would not have lived to write about them.)

John and I each drove a DUKW up the coast—the regular drivers were to join us later—and I well remember on the bleak day we approached Eureka that John stopped his DUKW and motioned for me to join him. He pointed out across Humboldt Bay and the sandspit separating bay from ocean to a sort of white froth on the horizon a couple of miles away, where an occasional geyser shot up.

"Some of those breakers must be 30 feet high—plunging. Look at them explode!" Then, in a matter-of-fact way, "That's where we're going to work."

Since I had never seen a wave before, much less the Pacific Ocean, I did not quit on the spot but accepted this proposal as a normal part of university research. Thirty-foot breakers sounded like a reasonable size for an ocean as large as the Pacific. Now it seems that even this modest description should at least have generated in my mind a picture of a tumbling wall of water higher than a two-story house breaking into a foaming mass that would compare favorably with the tumult below the spillway at Grand Coulee Dam, but it did not. Having spent my previous working life in mines and tunnels, I was not quite sure what a breaker was.

We set up an observation station at Table Bluff Lighthouse in Northern California, 100 feet (30 m) above the beach, and began systematically photographing the waves twice a day and noting their characteristics. Square frames covered with white canvas were anchored into the beach above high tide at 1,000-foot (330 m) intervals. These markers would give scale to the aerial photos and establish the survey lines that we would run out through the surf. On days when the surf was high, we would make simultaneous, radio-coordinated photos of the surf zone from the flying boat overhead and from the cliff. Then, when the breakers on the outer bar were low (meaning that most of them were under 15 feet, or 4.5 m, high), we would survey the bottom along lines extending out from the air markers. The object of all this was, of course, to establish the relationship between the waves and the underwater topography and give us data for future determination, by aerial photography alone, of the nature of the approaches to enemy-held beaches.

The method of surveying the underwater part of the beach was this: We would set up pairs of range boards 100 feet (30 m) apart perpendicular to the shoreline so that the DUKW driver could keep the craft on line as it moved slowly landward along the survey line. Down the beach 1,000 feet (300 m), at a point making a right angle with the range boards, a surveyor's transit was set up to measure readily the angle to the DUKW along the hypotenuse of the triangle. At frequent intervals a leadsman standing in the waist of the DUKW would call "mark" into the radio transmitter and heave a lead-weighted sounding line into water off the bow. As the DUKW passed the lead he would hold the line vertically and read off the depth of water beneath the trough of the wave. To this depth he would add the one-third of the estimated height of the wave about to break over him and call the total into the microphone.

The transit man, following the progress of the DUKW through the telescope, would, on hearing "mark" via the radio and seeing the lead splash, read the angle. An assistant at his side would record angle and depth. This process established depths at a series of points along a line, so that we could plot a profile of the sand surface beneath the waves. Because the beach constantly readjusts itself, there is no end to the work (see figure 44, page 228).

These jobs were divided in such a manner that John always ran the transit and I was the leadsman on these sorties through the surf. Somehow, in innocence and ignorance, I was persuaded that 15-foot (4.5 m) breakers smashing down on a 32-foot (9.7 m) tin boat was nothing to be disturbed about. In reality, of course, we often underestimated the height and unexpectedly encountered breakers over 20 feet (6 m) high. At these times, our friends on the beach offered many helpful suggestions by radio.

Perhaps when the Coast Guard from Humboldt Bay lifeboat station served notice that we were working at our own risk and could not count on their help if we got into trouble, I should have been more wary. They obviously were astonished that anyone would start out into what they considered to be a raging surf for any reason short of

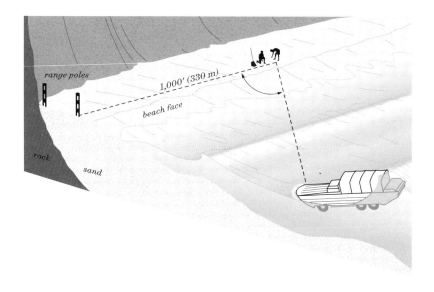

FIGURE 44: Once upon a time—surveying in the surf. The DUKW rides a breaker as it moves shoreward along a course marked by range poles. On the call of "mark" over the radio, the depth is taken by the DUKW, and the angle is measured from the beach to get the distance from shore. Today this is done from PWC equipped with a depth sounder and GPS.

emergency lifesaving. But they had the advantage of appreciating the risk, and several years had passed before it dawned on me that we were doing anything daring. We were the first, and one of the few, foolhardy enough to take this much interest in the sand beneath the winter surf on northern Pacific beaches.

There was many a close call when a DUKW would almost get sideways to a breaker or have the canvas cover ripped off and the supporting ribs caved in. DUKWs have no flotation compartments. If one went down, there was precious little chance—the outer breakers being more than half a mile from shore and the water temperature in the 40s (9 degrees C)—that either the driver or the leadsman would make it back alive. Even though we always wore life jackets, our heavy clothes and the wild turbulence would quickly have exhausted us. On one occasion, a breaker heaved the DUKW onto the beach face on its side, wheels pointing out to sea, throwing me clear. The next wave set it back on its wheels without damage. Somehow, no one was ever hurt. We were young and lucky, and in return for the risks we had the fun of challenging the breakers.

In order to run a line of soundings, it usually was necessary to get out through two major lines of breakers; often, we would smash head-on into half a dozen big breakers and be carried backward before a series of smaller waves would arrive and let us cross the bar. At each breaker, the DUKW collided with the mass of solid green water moving (relatively) at 20 knots; at the moment of impact, each tiny leak in the driver's cockpit was like a fire hose turned on the men inside. Often the glass in the windshield would crack or the canvas would rip; we nearly always lost the windshield wipers and ended up with the pumps furiously throwing out a stream of water the size of a man's leg. Though each wave washed over us, the DUKW would shudder and rise above the surface; with persistence we would get through.

At such times a good strong rip current was appreciated. When these currents existed, they carved channels through the bars, and the water would be too deep for all but the largest swell to break. If the DUKW could get in a rip, it had a much better chance of crossing the bar without serious pounding; besides, its speed was increased by that of the current. Amid waves breaking on all sides, these passages were hard to identify. We would station a lookout on the bluff who could see the whole surf zone and, by means of the walkie-talkie radios, guide the DUKW into the rip channels, much as aircraft pilots are talked through low clouds in ground-controlled approach. Or sometimes, on approaching the outer line of breakers, we would rest in the quiet zone between the bars and watch the big ones go over for a while—occasionally deciding that the sand could just as well be surveyed on another day. The drivers knew that any serious mistake would be their last one, and it was not unusual for them to quit for good when the wheels touched safely down on shore. Then we would talk it over and decide not to sally forth into such large breakers again. The next morning, the drivers would be on the job early, ready to go.

Surfing in a DUKW was great sport. I well remember bucking out through the surf just south of the Columbia River entrance, finally getting beyond the outermost breakers and then sitting there for nearly an hour getting up enough nerve to run the breakers. As a trough passed, we could look down a dark, watery valley that disappeared into the fog in each direction; then we would be lifted up on the next crest. From

Bascom in a DUKW, surfing on a smaller, 12-foot (4-m) plunging breaker during a beach survey in the 1940s. The high point is Bascom's head. Carmel, California, 1947. *University of California*

this temporary vantage point we could see a dozen more huge crests approaching, and looking landward see the back side of lines of frightening breakers. They were all about the same—nearly 20 feet (6 m) high. Finally, we would pick what we thought were slightly lower waves and make a run for it. Usually our judgment of height was wishful thinking, but for excitement it beats a roller coaster any day: full speed ahead at 6 knots until you are overtaken by a wall of water as high as a house and moving three times your speed. The trick is to time the run so that the biggest wave breaks just barely ahead of you; then, you can ride in atop the breaking crest, crossing the bar just in time to get beyond the reach of the next wave. It's a good idea, but it's hard to put into practice.

As the wave overtakes the craft, there is a sickening moment when the stern begins to lift rapidly and the driver fights to remain square with the waves (encountering a wave sideways would mean disaster). Then as speed picks up, the craft tilts forward at 30 degrees or so and

Personal Water Craft (PWC) bristle with instrumentation and software: echo sounder (fathometer), current meter, and GPS. The sediment profiles have accuracies of 4 inches (10 cm) at PWC speeds of 4–6 knots (8–11 km/h). Even these methods are being replaced by autonomous air, land, and sea vehicles. *Andrew Stevens/USGS*

buries its bow until there is green water across the windshield. The entire craft seems about to flip end-over-end and you think, *Why did I ever get myself in a place like this? What a fool to go to sea in a truck!*

But then the wave begins to pass under, and the buoyancy of the forward end lifts the bow until it is a level platform projecting out 10 to 15 feet (3–4 m) above the slick green water surface of the trough below. The forward wheels and axle hang down so that the rushing water of the breaker crest beats against them from behind and carries this awkward-looking truck-boat forward like a surfboard. Now you are flying, perched on a wave making 15 knots, water boiling on all sides—an exhilarating ride. Taking soundings has become second nature; you heave the lead, estimate the still-water level, call the depth.

Soon, the wave sets the craft into the quiet water and moves on, leaving the DUKW to continue at its own speed, still surveying. There is another line of breakers ahead, but these are only 10 feet (3 m) high and now seem tame. Inside at last, the DUKW rides easily and safely on the leading edge of a foamline sometimes head high. As the air escapes the foam, the mass loses height and only a foot or two of green water reaches the beach face. When through the inner breakers, the driver engages the wheels, and as the water shoals, the tires touch gently as weight is transferred gradually from the buoyancy of the hull to the truck's springs. Finally, motor roaring and gears grinding, the craft climbs out of the water and lurches to a stop at the backrush to mark the end of the run; then, up the beach face, a truck once more. With tire pressure controlled from the cab, a DUKW can operate in the softest sand. Happy to be safe ashore, the crew open the drain plugs to let out the water added to the bilges by the breakers. They check by radio with the transit party down on the beach to see if all the data were properly received and recorded. "OK? Now we'll run the next line a thousand feet down the beach."

Although this method of surveying in the surf may seem crude, we would often repeat lines to check on ourselves; even in the rough surf there was rarely a disagreement of more than a foot. Because an echo sounder will not work amid bubbles and turbulence, the old lead-line method was the only technique for obtaining such data.

After several years of almost daily surf operations, John and I wrote a thick pamphlet that became known as the *DUKW Report*. It went into considerable detail about our experiences with these wonderful vehicles, including the fact that we lost two in the surf and one that rolled off a cliff in Oregon and plunged 200 feet (60 m) into the sea. A result of that report about our operations in heavy surf was that the US Coast Guard—which had previously regarded these tin-hulled trucks with some suspicion—began using them for surfboats at their lifesaving stations.

The big breakers were so far from shore that we were never able to get very good pictures of the DUKWs in really large surf. Photographs of the surf zone at Table Bluff taken with a long-focal-length camera show the DUKW a mere speck in a breaker over half a mile from shore. In the summer we worked south, often in the Monterey Bay area of California. There, while training a new driver for the much rougher northern work, we obtained the photo on page 230 of a DUKW surfing a modest 12-foot (4 m) breaker.

After five years spent largely in observing and recording waves, studying and surveying beaches, and photographing the entire US Pacific coast from ground and air, I recorded the status of the coast as of 1950 in a three-volume tome, *Shoreline Atlas of the Pacific Coast of the United States*. Many years hence, when the forces of erosion have had time to accomplish notable changes in the coast, this book should form a valuable basis for comparison. *(Note: These pioneering studies are still referenced by today's researchers. Some surf-zone studies have repeated the same transit lines completed by Bascom more than seventy years ago.)*

SURFING SCIENCE

Captain James Cook seems to have been the first Westerner to observe and record the sport of surfing. In 1777, Cook watched a native paddling a small canoe in Tahiti, toward "a place where the swell began to take its rise." Then, while "watching its motion attentively," he "paddled before it with great quickness until it overtook him ... with sufficient force to carry his canoe before it, without passing underneath. The canoe was then carried along at the same swift rate as the wave, till it landed him on the beach." Most astonishing, the man then paddled back out to sea "in search of another swell."

Two years later Captain Cook was in Hawai'i, and one of his lieutenants wrote that the natives indulged in this curious amusement when the surf was at its utmost height. Then, after a pause that writer H. Arthur Klein calls the "dark days," surfing began to be revived in about 1900, primarily by George Freeth and Duke Kahanamoku. "The Duke" set up the first surfing club at Waikiki; Freeth restored the lost art of standing up on a surfboard, and in 1907 he brought the sport to Redondo Beach, California. With the addition of the fin in the 1930s, the surfboard's maneuverability was increased and allowed the board to be decreased in size. By the 1980s, the triple-finned "thruster" board allowed the down-wave speeds to be efficiently transferred into lateral forces. The smaller boards improved the surfer's weight-to-power ratio, decreased drag, unleashed the wave's energy, and revolutionized surfing. However, surfers on smaller boards were frequently submerged

Duke Kahanamoku on a spilling breaker in Hawai'i. He helped introduce the ancient Hawaiian sport of surfing to the world. Waikiki Beach, Hawai'i. *Alpha Historica/Alamy Stock Photo*

waist deep in water, no longer basking upon 13-foot (4-m) boards. The wetsuit was invented in the early 1950s to keep divers warm. Since then, cold-water and short-board surfers have created innovative designs that optimize flexibility, paddling, and upper-body warmth. Today, the sale of surfing wetsuits far outstrips diving wetsuits. Fins, easily transported boards, and wetsuits made surfing a global sport.

Initially, surfing was mainly a Pacific sport, although now there are surfers all over the world from Iceland to Fiji, Indonesia, and Tierra del Fuego. Warm, subtropical beaches that have a reasonably good chance of long-period swell are most favored. Surfers have, however, been observed to wait hour after hour for a wave in weather that would make a brass monkey shiver.

The trick of surfing, of course, is to get the board moving and the weight properly balanced so that the down-wave slope drag can take over the work of propulsion at the moment the wave passes beneath. Usually the objective is to ride in the tube or to go as far as possible on a single ride, preferably with a combination of speed and maneuvering. This requires an appropriately designed board, an ideal set of waves, and a lot of know-how derived from practice and experience. The board, which must be of suitable size for the rider and the conditions, should also be smooth and have the right wetted area as well as the proper fineness (length-to-width ratio) to maximize the dynamic supporting force while minimizing the friction and form drag. The surfers must contribute to this process by putting their weight at the right spot on the board to adjust the shape and size of the wetted area. For most, the complexities of waves, gravity, and balance are not easily mastered. It is rare for beginners to be able to stand on the first try.

The best waves for surfing are plunging breakers in which the end of the *curl* (or, breaking section) is progressing sideways along the wave front, somewhere between one and two times as fast as any point on the wave is approaching the beach. The fact that the breaker is plunging means that the surfer has a wider choice of slopes. Generally, the steeper the wave front at the point of contact with the board, the faster the ride.

Dr. Terry Hendricks was the physicist and California surfer who calculated that for very tubular waves, the optimal slope for surfing is about 57 degrees. This was found to agree with the positioning of top surfers measured on photographs taken at Hawai'i's Banzai Pipeline. These large, plunging breakers can move at more than 12 mph (19 km/h). The surfer, however, moves faster because the board is sliding sideways across the face of the wave (often just ahead of the sideways-moving curl). If that motion is half again as fast as the wave velocity, then the surfer is moving at 32 feet per second, or about 22 mph (35 km/h).

On some larger waves, rough measurements of the surfer's speed have yielded values of 25 to 30 mph (40–50 km/h). At these speeds, about 1 horsepower (750 Watts) is being dissipated in drag forces.

Some features of a good surfing area are:

1. It should frequently receive swell with periods of at least nine seconds and preferably longer.

2. The breakers should be of the plunging type so that the face of the wave is sufficiently steep to provide good board speed.

3. The lateral speed of breaking—the sideways extension of the curl along the crest—should be between one and two times the onshore speed of the wave.

4. Any wind should be light and preferably offshore.

5. The shape of the bottom should be fixed (by a rocky reef, for example, instead of sand) so that the favorable breaking condition is not changed by the waves. It should also have an even slope or a set of reefs that can maintain wave shape over a wide range of tidal heights and swell variability.

6. The beach face should absorb and dissipate the wave that reaches it, or the shoreline should be slanted so that no significant wave reflections enter the surfing area.

7. A deep channel, with an outflowing current, is desirable for paddling out beyond the breakers.

8. The bottom should not have dangerous features such as sharp rocks or jagged coral, and the shoreline should not be so rough and rocky that lost boards are damaged or surfers endangered.

9. An "indicator" break, farther offshore, can be a helpful warning of the approach of a set of large waves.

10. Convenient sighting points onshore are helpful in relocating the best takeoff position.

11. The area should be remote, to reduce the number of surfers (according to most surfers).

Few surfing spots have all of these features, but it is obvious that quite a number have enough to make riding the breakers an increasingly popular sport. Since the first publication of this book in 1964, surfing

Gerry Lopez, at Banzai Pipeline (on Oʻahu's North Shore), pioneered surfing big, hollow waves. He combined new surfboard designs with masterful skill on waves that were previously unridden. *Dan Merkel/ A-Frame* popularity has increased immensely. Countless magazines, movies, and lifestyles have created a worldwide industry based on the flow of energy, gifted by the winds to the waves that share their energy with surfers. Surfing is truly a renewable-energy sport.

Where are the best surfing areas? Probably they have not all been found or publicized; in any case it would be hard to get agreement on the order of listing them. There would, however, be no argument that the following areas are all near the top:

The north shore of O'ahu in the Hawaiian Islands, which faces into the open North Pacific Ocean, is rarely without large waves in the winter. It has several great surfing spots including the Banzai Pipeline, Sunset Beach, Waimea, and Hale'iwa. On the west coast of O'ahu, the surf is generally lower, but when it's up, the place to go is Mākaha and, on the south side, Ala Moana. Throughout the world there are great surf spots: Australia, Indonesia, Tahiti, Fiji, and South Africa are just a few of the excellent areas. The classic film *Endless Summer* and the more recent *Step Into Liquid* are travelogues of great surf spots around the globe.

Regardless of the size of the wave, the surfer needs to attain a speed close to the speed of the wave just before it breaks. This speed is the speed of the wave in shallow water: $C = \sqrt{gd}$. Because waves break in a water depth that is greater than their height (usually about 1.3 times); a bigger wave travels faster, and the surfer needs to paddle harder. The maximum speed a surfer can paddle limits the size of wave that can be caught. World-class swimmers swim at about 5 mph (8 km/h), and human-powered boats top out at about 20 mph (32 km/h); a surfer paddling a surfboard is somewhere in between the two speeds.

Now the question is: How big is a wave that travels at about 20 mph (32 km/h)? If we satisfy the shallow-water wave-speed equation, the answer is about 20 feet (6 m) high. The surfer does the work to get up to a maximum paddling speed, and then the wave takes over and hurls the board and surfer forward. At the same instant, inside the surfer's brain, a very rapid "risk-reward" series of thoughts occurs as the surfer attempts to balance the force of the wave and gravity while sliding forward inside a plunging wave crest. Most surfers never attempt to surf such large waves and cleverly avoid the "fear-of-death" dilemma of the surfing experience. The necessary balance, timing, strength, and perception skills to surf large waves over 15 feet (5 m) are immense. It is visceral, mesmerizing, and poorly conveyed in words.

In the early 1990s, technology changed surfing. The quest for larger and hence faster waves required new methods and equipment. To overcome the paddling limits of surfers, personal water craft (PWC) were used (e.g., Jet Ski, WaveRunner, Sea-Doo, etc.) to tow a surfer beyond normal paddling speeds. Unfortunately, the higher towing speed frequently caused the surfer to lose control on the steep face of the wave. The lack of control was overcome by installing foot straps as already common on windsurfing boards. In this fashion *strapped surfing,* also known as *tow-in surfing,* was pioneered by surfers such as Laird Hamilton, Dave Kalama, and others. Big-wave surfing was redefined. Today, hydrofoil (foil) surfers can glide above the wave, separated from the high-speed effects of rough water.

To catch a 60-foot (20-m) monster wave, the surfer must reach a speed approaching 35 mph (54 km/h). The biggest waves ever claimed to have been surfed are at Nazaré, Portugal. Nazaré's swell emanates from storms in the North Atlantic, and it is refracted by an offshore submarine canyon as it moves toward the headland at Nazaré. Garrett McNamara is credited with a 68-footer (24 m) there in 2011; Benjamin Sanchis had a 100-footer (30 m) in 2014; and in November of 2018, Brazil's Rodrigo Koxa was credited with riding a monster wave 80 feet (24 m) in height. Some of the surfers riding these waves will reach speeds in excess of 50 mph (80 km/h).

Wipeouts can be sudden, intense, and deadly. Free-falling from 100 feet (33 m) takes about 2.6 seconds. A surfer falling off a large wave (referred to as "going over the falls") will hit the water at nearly 30 mph (48 km/h) after falling just 30 feet (9 m). Falling from this height will take just over 1.3 seconds. Impacting the water at these speeds breaks bones and takes lives. Piloting the PWC in large surf is almost as dangerous as surfing. If you are still curious, try jumping out of your car from a bridge at 30 mph (48 km/h) into a river below—it hurts. Then, swim through the white-water river rapids to the shore. If this sounds compelling, you might be a future big-wave surfer—or crazy, or both.

Great storms create long-period waves and lesser shorter-period waves which travel at different speeds. These differences in arrival

Massive waves, spawned by storm winds in the North Atlantic, refract into Nazaré's submarine canyon and can peak up to 100-foot (30-m) monsters. They attract all manner of big-wave enthusiasts, athletes (Garrett McNamara in this shot), and spectators alike. Portugal. *Tó Mané/Getty Images*

times is called dispersion and is what allows wave forecasters to accurately predict the arrival of great surfing waves from afar.

Immense waves have been surfed at Jaws off the north shore of Maui and Mavericks off the coast of Half Moon Bay in California. One hundred miles offshore from San Diego, California, is the Cortes Bank, where waves of more than 60 feet (20 m) in height have been ridden by surfers. Wave heights are frequently disputed due to the difficulty of measuring such giant waves. It is truly a spectacular sight—go and see.

In Southern California, south of Point Conception, very large swell is rare because much of the coast faces toward the south, and all of it is protected to some extent by the offshore islands and underwater hills (seamounts). In addition, it is farther from the major swell-generating areas than the Hawaiian beaches. However, there are "windows" between the islands where the North Pacific swell leaks through. At that time, Rincon and Lunada Bay near Los Angeles, and Rights and Lefts

Kyle Thiermann rides a big wave at Mavericks near Half Moon Bay, California. The spot became a well-known big-wave break in the early 1990s. But riding big waves is perilous: Mark Foo, a professional surfer and big-wave expert, died here in 1994. *Fred Pompermayer*

as well as Cojo, near Point Conception, are all popular. In the summer, occasional long-period southern swell from storms south of the equator or hurricanes off Central America and Mexico create superlative waves at Windansea or Black's Beach at La Jolla or Malibu near Los Angeles. Then San Onofre at San Clemente, with its super-slow break, is a favorite for many.

The famous "Wedge" at Newport Beach, California, is less ridden with surfboards, but it gets a lot of bodysurfers, belly boarders, and knee boarders.

Northern California offers a cool climate for surfing. But for ardent surfers, Steamer Lane near Santa Cruz creates great enthusiasm. In Australia, Burleigh Heads, Dee Why, and Shark Island are very popular spots. Champion surfers hang out at K-108 off Peru; Fiji, Tahiti, Niijima Island off Japan; or Jeffreys Bay, Durban's Bay-of-Plenty, or Nahoon Reef in South Africa. Cloudbreak south of Namotu Island in Fiji is one

The Wedge (Newport Beach, California) is a spot where reflected waves (off the stone breakwater and the backwash down the steep beach) combine with the oncoming waves to produce unpredictable conclusions. *Benjamin C. Ginsberg*

of the world's most challenging waves. There are many more great spots that remain unnamed—these are carefully guarded secrets.

It is also possible to ride the surf far inland. In 1969, an artificial 2.5-acre (1-hectare) lagoon called Big Surf opened in Tempe, Arizona. Big Surf replicates a short section of an ocean beach with 5-foot-high (1.5-m) spilling breakers. From the point where the waves form in water about 9 feet (3 m) deep, the surfers can ride about 300 feet (90 m) to a beach with palm trees.

The wave generator was devised by Phil Dexter. Pumps steadily fill the reservoir; gates are raised that release some 70,000 gallons (270 m³) of water in about two seconds. As the water squirts out along the bottom of the reservoir, it is deflected upward by a concrete baffle in the bottom of the lagoon to form impulsively generated waves.

Several other artificial surf spots have been built over the years including in Germany, Japan, Surf Snowdonia in Wales, Point Mallard Park in Decatur, Alabama, and BSR Surf Ranch in Waco, Texas.

The Kelly Slater Wave Company provides plunging breakers with competition quality far from the ocean and regardless of tidal phase. Lemoore, California. *Chris Burkard*

A newer version of artificial surfing is called "flowboarding," where a high-powered pump jets water up an inclined ramp. Hundreds of these pumped systems have been installed around the world, including on cruise ships.

Recently the Kelly Slater Wave Company revolutionized wave making in Lemoore, California, more than 100 miles (160 km) from the ocean. Professional surfer Kelly Slater teamed with Adam Fincham (who has a PhD in geophysical fluid dynamics) and invented a large, moving, wedge-like structure that creates a perfect wave (a solitary wave) in an artificial lake a third of a mile (700 m) long. The wave shoals over a complexly shaped bottom that generates both gentle and barrel sections. Although expensive to design, build, and operate, Slater's Surf Ranch can produce an 8-foot (2.4-m) wave. Recently, while overlooking a famous surf spot (Black's Beach in California), a surfer from Chile described some of the waves back home as being similar to a "Kelly Wave."

Surfing has become a global sport and leisure activity with billions of US dollars spent on surfboards, wetsuits, surf apparel, advertising, and tourism. In 2020, surfing would have been featured as an Olympic sport for the first time. The surfing events were scheduled to take place in the ocean on a natural beach 40 miles (60 km) outside of Tokyo, Japan.

One Wave at a Time

Abandoned by waves, low tide leaves the sandbars bare. The sand, woven with rivulets and formed by retreating waters, is destined to be retextured by the returning tide. Slowly the Earth and Moon intertwine their gravitational forces, and the sea returns. In the distance, the swell eagerly wraps around a point. These shoaling waves rise, drawing nearer to the paddling surfer. The power of the wave lifts and thrusts the water forward; for an instant all the forces of heaven and Earth are balanced. Then, tethered to the Earth by gravity, the wave is pulled downward, spilling forward and crashing into turbulent energy.

The tide continues to rise. It enables the shoaling waves to reach new areas of the bar. Small sand waves get massaged, one wave at a time. The speeding water creates viscous forces, making viscosity king, and overwhelming both buoyancy and gravity. The mix of sand and water is resuspended in billows, each grain traveling its chaotic path. The wave passes and the flow of energy momentarily rests. Gravity regains control as each grain of sand is repositioned. Every wave repeats this process. By the time the tide recedes again, the entire beach will have been redrawn as nature intended.

But not yet. When everything is just right, I decide to paddle out. A large wave arrives and I inhale deeply. With momentum I push the nose of the board downward. I stop breathing, close my eyes, and submerge in front of the advancing wave. For a fleeting moment all is silent, and then I hear whooshing, and the wave is everywhere above and around me. The rushing water's forces attempt to separate me from my board, but I grasp tighter. After the crest of the wave has passed overhead, with the foam still above me, I open my eyes and behold all is blurred in the subdued light. The turbulent forces decrease, and my buoyancy assumes command. The sky returns, the water rushes off my body, my vision clears, and without delay I paddle out toward my awaiting buddies in the lineup.

Beyond the breakers, I rest, looking seaward and poised for the arrival of the next wave. It could be perfect; neither too steep nor too small. I rotate as one with my board, toward the land. My heart begins to pound, and I begin to paddle.

"Paddle, go hard, it's yours!" shout my buddies.

In an instant I pop up. Unknowingly, I am in command of gravity, buoyancy, and viscosity. I am positioned on the face of the wave: gravity pulls my board downward as the wave accelerates everything forward, toward the wave's trough. At maximum speed, maximum kinetic energy, I start my bottom turn—my head turns, and I twist my hips and bend my knees. I aggressively change direction and head back toward the wave crest. The movement makes me heavy on my board but my bent knees and thighs absorb it and force my feet harder into the board. I leave my maximum kinetic energy behind. As the crest is reached, the nose of the board is now sloping skyward at the point of maximum potential energy. And again, with a snap and twist of my body, I am off the lip. I flow back downward propelled toward the shore. This dynamic exchange of energy is repeated until the wave's energy or my body is exhausted. I kick out of the wave and start to paddle out.

The sky has dimmed. The breeze has freshened and formed ripples everywhere. The next wave is slightly different, the tide has changed, the flow over the bar altered, the longshore current has relaxed, the rip is stronger, and each grain of sand has moved. It will ever be so for each and every wave. Surfing is an addiction to a rush of air, to the glide on water, and to the uncentered flow of wave energy. It is a tribal codependence upon the power that moves the stars. Go and see. – KM

When I have seen the hungry ocean gain
Advantage on the kingdom of the shore,
And the firm soil win of the watery main...
– William Shakespeare, Sonnet LXIV

10 Beaches: Where the Surf Meets the Sediment

Along the boundary between land and sea, the solid underlying rock is covered with a layer of rock fragments. These fragments range in size from fine sand to large cobbles, in thickness from a few inches to hundreds of feet, in colors and shades from clear white to opaque black. These are beach materials. Every coastal dweller in the world is quite sure what a beach is like. Yet if you were to ask, you would find totally different opinions, and mostly derived from local experience. Sandy beaches compose more than 30 percent of the world's ice-free shorelines.

BEACH MATERIALS

The open-sea beaches that border much of the United States, from Cape Cod south along Jersey and the Carolinas to Florida, and along

PREVIOUS SPREAD: Beaches form from materials that are broken up by waves and raised and lowered by the tides. These black grains of sand are volcanic in origin and coarse enough to allow the water to seep into the beach face. Waiʻānapanapa State Park, Hawaiʻi.
Matt Anderson/Getty Images

the California coast south of Point Conception, are for the most part composed of coarse, light-colored sand produced by the weathering of granitic rocks into their two main constituents, quartz and feldspar. Generally, these beaches are steep-faced and coarse-grained. Because they contain our most popular beach resorts, many Americans tend to think that they fairly represent the world's beaches. But hundreds of miles of beach along the Oregon and Washington coast are quite different. There the sand is fine-grained and dark gray-green in color, derived from the weathering of basalt, which forms beaches that are wide, flat, and often hard as a racetrack. Much of the Florida coast is equally hard and fine-grained, but it comes from the disintegration of coral. On the other hand, the beach at Cannes in southern France is largely composed of uncomfortable pebbles, and much of the English coast is lined with small, flat stones called "shingle." In fact, the word "beach" seems to have been the ancient word for "a shingle shore."

Many beaches of Labrador and Argentina are composed of large cobbles. Those of Baja California are composed of two materials, a flat sandy portion that is exposed only at low tide, while above and behind the sand great cobble steps called "ramparts" rise to a height of 30 feet (10 m) or more. In Tahiti, if you live on the windward side of the island, you think it is natural for a beach to be made of black volcanic sand. But if you live on the other side, where the wide coral reef furnishes the beach material, it seems reasonable for beaches to be blindingly white. In fact, beaches can be made of nearly any material that is present in quantity; rock fragments are not necessarily required. At Fort Bragg, California, a small pocket beach once consisted entirely of old tin cans washed in from the city's nearby oceanic dump and arranged by the waves into the usual beach forms, as though to prove that the laws of beach physics cover all possibilities.

Thus, although beaches vary widely in appearance and composition, the principles that govern their behavior are the same, and for convenience here, all beach materials will be called "sand." Table 10.1 on page 250 lists the sizes of the particles and may help the reader to visualize the beaches being discussed relative to those within one's own experience. Several other factors, including the shape and density of the particles, are of interest, but they are of secondary importance.

A beach responds with great sensitivity to the forces that act upon it—waves, currents, winds. It is a depository of material in transit, either alongshore or off- and onshore. The important thought in the definition is that of motion, for beaches are ever-changing, restless armies of sand particles, always on the move. Most sand movement occurs underwater, the result of waves and wave-caused currents that organize the particles into familiar forms. But the motion of a beach created by the waves, even when huge quantities of sand shift in a single day, may not be noticed by a casual observer. The short-term changes are usually imperceptible.

TABLE 10.1 Sizes of Beach Materials
(US Army Corps of Engineers Standard from Coastal Engineering Part III-1-8)

Material		**Size** [millimeters]
Boulders		Larger than 200 (over 8 inches)
Cobbles		76 to 200 (3 to 8 inches)
Gravel (includes shingles and pebbles)	Coarse	19 to 76 (0.75 to 3 inches)
	Fine	5 to 19 (0.2 to 0.75 inches)
Sand	Coarse	2 to 5 (0.08 to 0.2 inches)
	Medium	0.4 to 2 (0.016 to 0.08 inches)
	Fine	0.07 to 0.4 (0.003 to 0.016 inches)
Silt or clay		Less than 0.074 (barely visible to naked eye)

Watch the waves break on a sandy beach; the water runs up the beach face a short way; some of it sinks in, the rest slides back down as the backwash. The moving water carries a film of sand in each direction, and the question is, What is the net effect? Is sand being added to or subtracted from the beach face?

For any small number of waves, no one can give a positive answer; each wave is slightly different in height or velocity and may either add or take away a few grains of sand. But overnight, or after a week, the net effect of the waves may be easily observable. Now you notice that a rock is covered (or uncovered) by sand; you see a small vertical cliff cut into the berm or a newly added ridge of sand along the beach face;

Sand grain sizes. Fine (top), medium (middle), and coarse (bottom).
Kim McCoy

a little way offshore the waves break in a different place, indicating that the bar has shifted. The sand feels different beneath your feet—a new layer of sand, not yet compacted by the waves, is soft to walk on. These are evidence of beach motion; whenever and wherever there are waves, there is constant shifting, constant readjustment.

This chapter deals only with the offshore-onshore motion of sand. *Littoral transport*, or the flow of a stream of tumbling particles of sand alongshore under the influence of wave-caused currents, is described in Chapter Eleven.

Sand Motion

The principal beach forms are shown in figure 45 (page 254), which is a generalized profile of the conditions that prevail in winter and in

LA JOLLA SHORES IN WINTER The berm level has been lowered more than 6 feet (2 m) and moved to an offshore bar by the larger and steeper winter waves. La Jolla, California. *Kim McCoy*

summer on many beaches exposed to the ocean. Remember that our definition of beaches includes all the sand in motion above and below water out to a depth of about 30 feet (10 m). Above water, there is usually a nearly horizontal terrace of sand brought ashore by the waves: the berm. Below water are elongated mounds of sand that parallel the beach called bars, or sometimes longshore bars.

In summer, the berm is low and wide. To the layperson it is the beach—the observable sand on which beachgoers sunbathe and frolic. At that time, the underwater profile is likely to be smooth and barless.

In winter the berm is higher and narrower, as most of the sand moves underwater to create the offshore bars. The reason for the shift is the change in wave action. The large waves that come from winter storms cut the berm back; the small waves of summer replace it again. If the amount of sand involved is constant, as it is on a beach between

LA JOLLA SHORES IN SUMMER A few months later, the summer berm is back at the base of the stairs, moved by smaller, more gentle waves. This seasonal transit of sand is found on many beaches. La Jolla, California. *Kim McCoy*

two rocky headlands, the entire beach motion is merely an exchange of sand between berm and bar.

Therefore, the study of beaches that are closed systems is concerned principally with the questions of why the sand moves in each direction, which waves are responsible, and what shapes and slopes the sand takes. There is a rather delicate balance between the forces that tend to

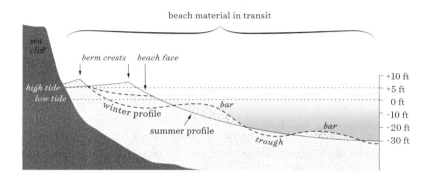

FIGURE 45: Generalized profile through an intermediate-slope beach showing seasonal changes in the distribution of the sand.

bring sand ashore and those that move it seaward. The position of the main mass of sand is a measure of which force is dominant.

The basic mechanism is simply the lifting of the individual sand grains from the bottom by the turbulence accompanying the passage of a wave. A sand grain weighs little because it is lighter underwater than in air (buoyed by the weight of the water it displaces) and not much energy is required to lift it. Moreover, because of the turbulence and viscosity of the water, the grains settle slowly. While grains are in suspension, or falling freely, currents of very low velocity can move them sideways. Each time a sand grain is lifted, it lands in a slightly different location. Uncounted millions of sand grains are picked up and relocated by every wave, and the beach constantly shifts position. They need not move very far each time, for there are some 8,000 waves a day. Sand grains that move a tenth of an inch (2.5 mm) per wave could migrate 70 feet (20 m) in a day. Below the

Sediments suspended in a steep wave an instant before plunging into turbulence. *Danny Sepkowski*

wave, the pressure of each passing wave crest squeezes water through the gaps between grains of sand. Of course, all waves do not have the same effect, and the currents may change direction. Hence, it is difficult to say whether the sand is moving to or from shore at any moment.

The key to the relation between waves and sand motion is the large change in the beach between winter and summer. Clearly, there is a difference in the kind of waves, but what is it? In winter the waves are large and the surf is rough; suspended sand can be seen boiling up behind a breaking wave. Energy is being expended on the beach at a higher rate than in summer. This rate of delivery of energy is most conveniently described in terms of wave steepness—the ratio of wave height to wave length, commonly written $\frac{H}{L}$.

For example, a 6-foot (2-m) wave 600 feet (183 m) long has a steepness of $\frac{6}{600}$ or 0.01. If the wave length is only 200 feet (61 m), 6-foot

(2-m) waves have a steepness of 0.03. Thus, wave steepness increases either with an increase in height or a decrease in length. In wave-channel experiments, J. W. Johnson of the University of California at Berkeley was able to show that when the wave steepness is greater than 0.03, bars always formed (starting with a barless beach profile). If the steepness was less than 0.025, bars never formed in the model tank. On a real beach the values are probably different, but the essential idea is the same. On the beach the waves are highly variable in both height and period, in contrast to those in the model tank, which are all precisely the same. Moreover, it is difficult to assess the effect of sand size, which, if scaled down to conform to the rest of the model, would be too small to react properly.

The effect of wave steepness seems to be as follows: When the waves approaching the beach are small (or the wave length is long) the sand on the bottom moves shoreward with the orbital currents. These low-steepness waves pick the sand up, move it forward, and set it down. Although the orbiting water returns seaward an equal distance, the sand it carries is now more likely to be dragged along the bottom. Friction against other sand grains and the existence of a laminar, or nonturbulent, flow region at the bottom keep the sand from moving quite as far as the water does and thus from completing the orbit. Consequently, the net motion of sand is landward when the steepness is small.

When relatively large waves follow close upon one another, an entirely different set of circumstances exists. Now there is general turbulence in the surf zone, which keeps the sand in suspension, particularly in shallower water. The mass transport of water by the high waves is greater, and when they break, substantial waves of translation are generated. The result of these effects is that there is a general flow of water shoreward along the surface. Because the waves are relatively close together, the berm remains saturated, and relatively little of the water traveling up and down the beach face sinks into the sand. The shoreward-moving water carries a load of suspended sand particles, and when the waves rush up the beach face, their leading edges surmount the crest of the berm and deposit their sand atop the berm, raising its height. The remainder of the water rushes back down the beach face, picking up a thin layer of sand as it goes. This sandy suspension becomes involved at the bottom with the seaward-flowing currents, which must, of course, balance the landward-moving water at the surface. These currents move the sand seaward until they reach the breaker zone, where the landward-flowing currents are generated. There they deposit their load to form a bar. (Note that this current,

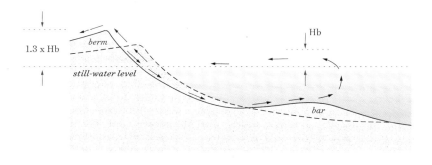

FIGURE 46: The circulation of water in surf where steep waves transform the bottom from the dashed profile to the solid profile by moving sand from berm to bar. The height of a berm is a complex balance of run-up, swash over-topping, and tidal elevations, generally about 1.3 times the height of the breaking wave (Hb).

while strong enough to influence suspended sand, could not be detected by a swimmer.) Thus, we have an explanation of how the steep winter surf can build the berm higher while cutting it back, and how bars are formed by the erosion of the berm (see figure 46).

The difficulty in determining whether a berm is retreating or advancing at any moment comes from wave variability—the difference in height and length from wave to wave. Suppose that the waves arriving at a beach all have about the same period and wave length but the height varies as it does in the examples of figure 38 (page 197). The small waves would bring sand ashore; the large ones would take it away. And that is the way with beaches. The sand is constantly shifting in accordance with complicated and variable water motion. The profile of the sand itself is a rough analogue solution to the question: Is the average wave steepness above or below 0.03—is there a bar or berm being formed?

Berms and Bars

Now, equipped with a general understanding of the mechanics of sand migration in the surf zone, we can examine more perceptively the beach forms that are produced. To do this, one takes a series of profiles of a beach over a period of time and examines the changes. Over the same period, the waves reaching the beach are observed and recorded. The idea is to correlate the beach changes with some specific quality of the waves—but it is not easy to obtain the information or to make sense of it. Eventually, though, the persistent researcher does end up with an accumulation of data and a feel for the way that waves and beaches interact.

The Waves field party, which started in the 1940s, eventually surveyed beaches at some forty Pacific coast locations, repeating profiles at some of them dozens of times in many kinds of weather conditions and in all seasons of the year. On each visit we surveyed three lines 1,000 feet (300 m) apart, extending from the dunes to 30 feet (9 m) below sea level to ensure obtaining a representative profile. In the course of five years, about 500 profiles were made and 600 sand samples were taken. Figure 47 compares the profiles of flat, intermediate, and steep beaches. Because beaches are very irregular in slope, the words "steep" and "flat" are relative. As used here, a flat beach is one on which the water is less than 10 feet deep 1,000 feet (3 m/300 m) seaward of the zero tide level (or, mean lower low water), whereas a steep beach is more than 30 feet deep 1,000 feet (10 m/300 m) out from a similar point. Note the difference in vertical exaggeration between the figures.

Each of these beach profiles was selected because it is of special interest. Omaha Beach, a principal landing point in the Allied invasion of Europe in 1944, is a flat beach. The tidal range there is about 15 feet (5 m). At low tide, the German defenders could plant the beach tank traps and landing craft obstacles, which were such serious problems to Allied D-Day operations.

Leadbetter Spit, typical of 100 miles (160 km) of beach north and south of the Columbia River, was intensively studied because of its similarity to the Japanese beaches. After having been in and out through the surf at Leadbetter many times, we reached the conclusion that

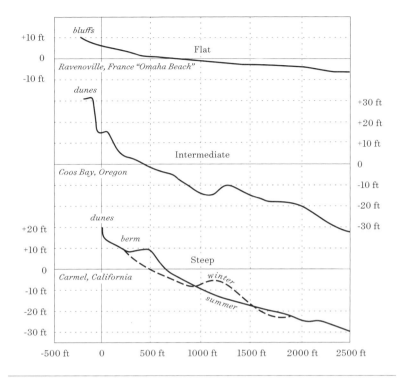

FIGURE 47: Beaches with flat, intermediate, and steep profiles. The top profile from France is continual and gradual. Coos Bay and Carmel have similar slopes for part of the year, but in summer, Carmel's berm descends rapidly into deeper water, presenting a much steeper profile.

except in very low surf, the attempt to land on Japan would have been an amphibious disaster.

Coos Bay, Oregon, has a pronounced bar, a clear winter berm, and a steep-fronted dune line caused by a violent storm that had cut deep into the semipermanent dunes. The huge bars at Table Bluff have an attitude of violence about them that they well deserve—or so they look to me.

The steep beaches, particularly in Carmel, California, demonstrate the extreme. There it is not unusual for waves to break with great violence directly on the beach face, impelling a thick uprush or surge of water up and across the beach that can be dangerous to both swimmers and landing craft. Since this book was first published, the intensely turbulent surf at Fort Ord (near Carmel) has been studied many times.

The Fort Ord Swash Machine

I was "sucked into" Fort Ord research in the 1980s. As Fort Ord's violent waves break, the wave's energy shakes the land, spurts water into the air, and sends sounds hissing and rumbling back into the depths of the sea. The intrigue of these sounds and the unknown path of the energy beckoned scientists, despite the dangers.

A challenging plan was devised to pass instrumentation cables through the surf zone. In a daring move, the captain of the research vessel Acania *edged it toward the dangerous breaker line at Fort Ord. At the moment the captain began to fear for his life, crew, vessel, and sanity, the engines were reversed. Simultaneously, I jumped overboard, knife in hand. Most waves were over 13 feet (4 m) in height; a polypropylene line was tied around my waist and was paid out from the ship as I swam. The intention was to swim directly to shore but the fast longshore current took control.*

Every wave became more humbling than the previous and the longshore drift intensified (faster than I could swim). Although a strong swimmer, I was severely churned by the surf while still attached to the polypropylene line. I was in the middle of a tug-of-war between the force of the waves and the propellers of the Acania. *Eventually, the longshore current handed me over to the swash zone. The swash rolled me up and down the steep (1:5) beach face. I had entered the spin cycle of the (s)wash machine. I never cut the polypropylene line.*

Exhausted, with the unused knife still in hand and somehow still alive, I exited the sea more than 1,000 feet (300 m) toward Monterey. Once ashore, my end of the polypropylene line was attached to a winch and the heavy steel-clad instrument cables were pulled from the Acania.

The experiments at Fort Ord were a great success, and the memory of the grueling swim remains as a life-threatening, humbling, character-building experience. – KM

Much of the early beach profile data came from Carmel, California. The beach is a closed system, because it is protected by headlands at each end and by a deep reef offshore. Ten-foot (3-m) breakers are not unusual. In 1946, a careful study was made on that beach in an attempt to keep a "budget" of the sand position—that is, to know where all the sand was, all the time. Figure 48 shows the growth of the berm; in the five months between April and September it widened by more than 200 feet (60 m). By December the first large storm had caused the beach face to retreat substantially, and by February it was almost back to the point of beginning. It was possible to detect berm growth by making precise surveys hour by hour, and we observed growth rates of more than 6 feet (2 m) a day. When storms start to erode the berm,

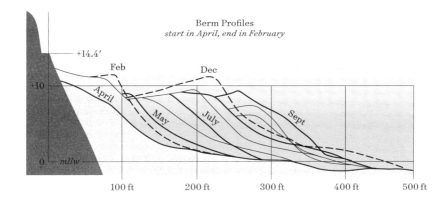

Berm Profiles
start in April, end in February

FIGURE 48: The berm in Carmel, California, builds 200 feet (60 m) seaward during summer when the waves are small (September profile). Then it retreats almost to the vanishing point in the large waves of winter storms.

particularly during neap tides, it is not unusual for a steep sandy cliff or scarp to form at the seaward edge of the berm. A vertical scarp 5 feet (1.5 m) high was seen on the Oregon coast, cut overnight by a short and violent storm. On the next survey, two weeks later, small waves had replaced most of the eroded material and only the upper 1 foot (30 cm) of scarp remained. The beach face, of course, is immediately seaward of this scarp, and its slope is actually flattened by the pounding waves.

Since berms are formed by wave action, it is not surprising that the height of the crest of the berm is a function of the height of the waves. Experiments in the 1940s by R. A. Bagnold of England in a wave channel demonstrated that the height of the berm above sea level is 1.3 times the height of the (deep-water) waves that formed it. A similar

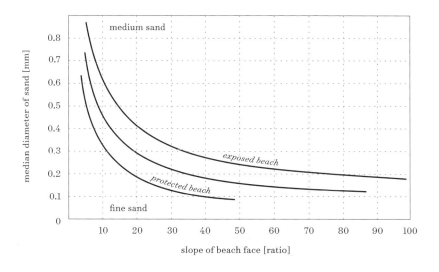

FIGURE 49: The relationship between sand size and beach-face slope at the mid-tide zone on exposed and protected beaches. The upper curve is the minimum probable slope; the lower is the maximum.

relationship also exists in nature, but it is difficult to confirm because sea level constantly changes with the tides, because refraction influences the amount of deep-water wave energy that reaches any beach, and because every wave is different. However, it seems likely that on ocean beaches the height of the berm is about equal to 1.3 times the significant height of the deep-water waves multiplied by the refraction coefficient. For example, the berm on the continuous beach rimming the gently curved shore of Monterey Bay is 16 feet (5 m) above low water at the exposed Fort Ord section and decreases gradually in height toward protected Monterey harbor, where it is 6 feet (2 m) lower.

Even a casual beach-watcher soon notices that the slope of the beach face and the size of the sand are somehow related. Steep beach, coarse sand; flat beach, fine sand. But determining more precisely what the relationship is took quite a while. There are lots of places on a beach to take sand samples and to measure slope, and on the same beach the results vary greatly from place to place. A good deal of sampling and measuring was done before a reference point was selected. If the sand sample is taken on the beach face in the zone subjected to wave action at mid-tide and the slope is measured at the same place, consistent

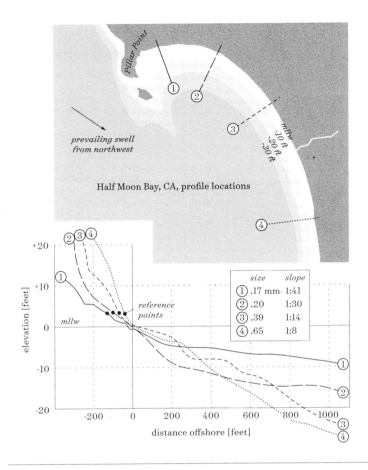

FIGURE 50: The effect of a protecting headland on beach slope and sand size.

results are obtained. Figure 49 shows the relationship for exposed and protected beaches. Here again the effects of wave energy cause variations. If the berm is retreating before the pounding of the waves, the slope is less steep than if sand is being added.

A more important effect of wave action on sand size and slope is illustrated in figure 50, which shows four beach profiles made along the continuous slope of the beach face at Half Moon Bay, California. There, Pillar Point completely protects the beach at profile number one from the prevailing northwest swell. Profile number four is exposed; between these extremes are two beaches of intermediate slope. In the protected zone behind the point, the beach is flat and the sand fine, but toward the south the beaches grow steeper and the sand coarser. This demonstrates how beaches adjust to the wave environment.

Longshore bars (or beach bars) are underwater ridges in the beach material that parallel the shoreline. Often more than one bar is present, the number depending on the size of the waves, the bottom slope, and the tide range. We have discussed already how these offshore features are formed by currents that flow when steep storm waves arrive. When formed, bars have a pronounced effect on the waves. Because they are abrupt shoals, they tend to act as filters, causing all waves above a certain size to break at one spot (instead of breaking over a wide zone as they would on an even slope). Moreover, because bars slope steeply to seaward, they tend to make the breakers plunge and release the energy suddenly.

A substantial range of tide (of 5 feet or 1.6 m, or more) tends to create two sets of bars (see photo page 45). Because the tide curve is sinusoidal, sea level lingers near high tide and low tide for extended periods but changes from one to the other with relative rapidity. Bars can develop at two levels corresponding to high and low tide, but the sand cannot shift from one to the other during the change. Surveys have determined the depths of bars on many beaches under various conditions of storm and calm. In tabulating the results of twenty-nine surveys of exposed beaches, we discovered that all had at least one bar and that those with average underwater slopes of less than 1:75 had three bars. The top of the inner bar was usually about 1 foot (30 cm) below mean lower low water. The average depth of the top of the second bar was 7.5 feet (2.3 m), and that of the third, or outermost, bar was 13 feet (4 m). Thus, the difference in depth between the bars is about 6 feet (1.8 m), which is the usual range of the tide on that particular coast.

When the tide is low, large breakers break first on the outermost bar; then they re-form into waves and break again on the second bar. At high tide, the outermost bar may be too deep to cause the waves to break, and they pass over to break on the two inner bars. The deepest bars formed by great storm waves remain unchanged through months of calm weather because small waves do not reach deep enough to rearrange the sand. Even after the most violent storms, we never found bars with crests deeper than 20 feet (6 m) below mean lower low water, and we concluded that this was their probable maximum depth.

Like other beach features, bars have greater relief when they are made of coarser sand. That is, the troughs between bars are deeper, and the slopes are steeper. Various attempts to correlate bar spacing with wave length or to find a simple relationship between depth of bars and troughs have been made. Because longshore currents often flow in the troughs, these may be scoured deeper or filled in, independent of wave action.

Bars have been observed on beaches ranging in size from model tanks to lakes and oceans, and subject to corresponding wave actions. As many as five have been observed on a single beach, and substantially unbroken bars 20 miles (30 km) long have been observed on Leadbetter Spit, Washington. But although they have been described here as if they were continuous parallel beach features, often they are very irregular. When wave direction changes, the bars begin to shift position; if the waves quiet down before the shift is complete, the result might be an indescribably ragged arrangement of sand that will remain until the next storm produces order again.

The question of the maximum depth at which the bottom material can be moved by wave action is not settled, in spite of the arbitrary 30 feet (10 m) below low tide used herein. For example, a violent storm in Madras, India, cast up on shore a quantity of pig lead that came from a vessel wrecked more than a mile offshore. Shingle and chalk ballast dumped overboard by sailing ships in water more than 60 feet (18 m) deep and more than 7 miles (11 km) from shore was brought ashore at Sunderland, England, by wave action. And captains of ships passing over Nantucket shoals, where the depth is 75 feet (23 m) and more, report that storm waves breaking over the ship leave sand on deck. What is the mechanism of transport? No one knows exactly.

The delicate balance of berms and bars can be easily upset. Coastal structures, artificial reefs, and dredging are only a few human activities that create imbalances. Changes on land including types of vegetation, erosion control, sand mining, housing construction, roads, and wind patterns can cause disruptions in the coastal dune stability that affect the berm and bar dynamics. Artificial reefs, when properly placed, can diminish wave energy, protect shorelines, and augment the formation of berms and bars.

Minor Beach Features

There are several beach features that seem to have no great geological or engineering importance, but they add interest to the study of beaches. These include *swash marks, backwash marks, rills, steps, cusps, domes, pinholes,* and *ripple marks.* A wonderful time to observe these features is early in the morning, especially after a high tide. Often the air is still, and a pleasant light fills the sky. The beach is clean, the night's waves having erased the human marks of the previous day. Then, with the sand surface free of confusion, beach-watching is most rewarding.

The flow of a thin sheet of water up the beach face that follows the final breaking of each wave on the shore is called the *uprush,* or sometimes the *swash.* At its upper edge, just before its energy is spent in the upward motion, the swash is only a film of water less than a quarter of an inch deep. Immediately ahead of the moving water is a line of sand particles, usually a little larger than most of those on the beach face, bulldozed along by the edge of the swash. When the momentum of the moving water is spent in upward motion, the uprush stops. Part of the water sinks into the sand and part of it slides backward down the beach as the backrush. As it does so, a thin line of sand grains is left to mark its maximum upward reach. These are *swash marks.*

Swash marks are a sort of scorecard of the reach of a succession of waves. As the general level of surf rises and falls with surf beat, carrying on it large and small waves, you can watch the marks change position. As the level recedes and successively smaller waves create uprushes, the swash marks make a pattern like that shown in the photo on page 267. But as the amplitude of the waves increases, suddenly one larger wave will erase all the previous marks, using their sand grains to make its own mark. On a receding tide the highest marks remain until the next incoming tide, when a new series, erasing and replacing, climbs up the beach face.

The part of the swashing water that does not sink into the sand but runs back down the beach face (the *backrush*) often creates a diamond-shaped

SWASH MARKS Swash marks remain on the beach face after the uprush of each wave has ended. *Kim McCoy*

pattern of backwash marks. Somehow, the moving water flows in a manner that creates tiny crisscrossed valleys about one-fourth of an inch deep. Usually, these backwash diamonds are about 6 inches (15 cm) long, and the long axis is always oriented perpendicular to the shoreline. The diamonds are most likely to be seen on beaches of intermediate steepness and moderately coarse sand. Why a flat sheet of moving water should make tiny gullies in diamond form remains a mystery.

When the tide retreats, the water left on the higher part of the beach will seep out through the sand and flow down the beach face. The pattern of drainage of these tiny river systems is known as rill marks, which look something like plant stems that branch outward toward the sea and, when seen in ancient rocks, have been repeatedly mistaken for fossil plant remains. Note that the pattern spreads outward as it descends, like a delta system rather than as a river system in which tributaries join to enlarge the main stream.

BACKWASH MARKS Diamond-shaped backwash marks created by a thin sheet of water rushing back down the beach face. *Kim McCoy*

When found in sandstones of other geologic periods, swash marks, rill marks, and backwash marks are not only excellent evidence that the sand once was the beach of a prehistoric sea, they also point the direction of the sea and indicate the range of tide.

As the backrushing water slides down the surface of the beach, its speed increases and the surface sand grains are lifted into suspension. The momentum of this turbid sheet of water and sand carries it below the general level of the sea, and a small wave, usually less than a foot high, called the backrush breaker, curls over it. The result is a turbulent sandy swirl whose effect is to lift the sand grains and keep them in suspension, to be carried up the beach face again by the next uprush. But the larger sand grains may settle to the bottom, where they roll back and forth as each wave passes, occasionally being lifted and dropped again by unusually violent water motion. The net effect of this constant shifting is segregation according to size in which the larger particles

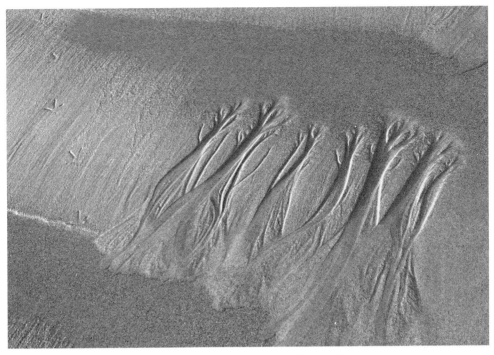

RILL MARKS Rill marks are grooves and channels made by small streams of water seeping through the sand, down the beach face, and following the outgoing tide. *Christina Speed*

move steadily downward. A little below the level of the lowest backrush breaker, these larger sand grains reach a depth at which they can no longer be moved upward by most of the waves. The result is a step-like deposit whose upper surface is the continuation of the beach face and whose outer surface is the angle of repose of the sand. This low-tide step, which is usually about 1 foot (30 cm) high, may be hard to see because of the turbulence, but it is often encountered with momentary alarm by the bare feet of waders who sink into the soft, coarse sand or step off its abrupt edge.

Cusps are evenly spaced, crescent-shaped depressions concave to seaward that are built by wave action on the seaward edge of the berm (see photo on page 109). Of all the curiosities of the shore, these are surely the most puzzling. Dozens of explanations have been given for their formation, although standing edge wave theory seems to be in general acceptance. (Beach cusp theory gets complex and we will leave

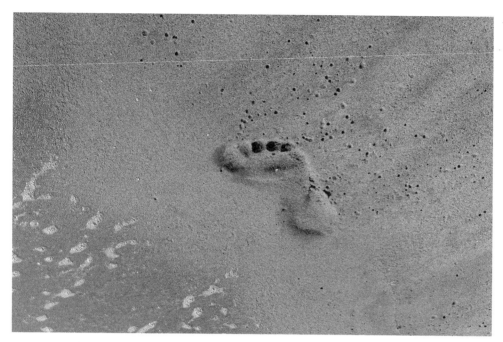

PINHOLES Pinholes are formed near the upper end of the swash
when water sinking vertically into the sand forces out the air between
dry grains. *Kim McCoy*

its theory for others with more time for in-depth explanations.) Cusps
vary greatly in shape and have been observed in beaches made of fine
sand and large cobbles; they occur equally in protected bays and on
exposed beaches. Cusps have been made in the laboratory with lengths
of 6 to 9 inches (15–23 cm); they have been measured at San Simeon,
California (14 feet, or 4.2 m), at Fort Ord (90 feet, or 27 m), and at Table
Bluff (average of nine cusps: 1,180 feet, or 360 m).

Like other beach features, cusps have more striking relief and are
less regular on exposed beaches made of coarse sand than in protected
bays, where the precision of cusps sculpted in fine sand on a flat beach
is a thing at which to marvel. They seem to develop at wave steepnesses
between erosion and deposition. In fact, cusp shapes are the result of
a dynamic balance of changing wave forces, beach slopes, sand grain
sizes, and beach swash zone porosity. An observer can stand for half
an hour watching wave action in cusps and remain fascinated by how
the cusps persist.

Researchers generally agree that (1) conditions are best for cusp formation if the waves approach exactly parallel to the shore, and are unconfused by local currents and winds; (2) some original irregularity in the beach is necessary to start them forming; and (3) the spacing of the cusps is related to wave height. Why cusps form and the relation between wave height, run-up, beach slope, and cusp length is complex and continues to be studied by beach researchers. Cusps perform a role in beach dynamics; when we upset the delicate balances of coastal energy flow, we change the beach morphology.

In the dry sand on the beach face above the usual swash, the spaces between the sand grains are filled with air. In this area, the sinking water of an unusually long uprush will cause sand *domes* and *pinholes* to form. This happens when the thin layer of uprush water sinks vertically and displaces the air. The air then migrates upward and emerges from pinholes in the sand surface as a chain of bubbles. The water swashing across the pinholed surface in subsequent uprushes sinks down through the holes and smooths them into tiny, temporary funnels about an eighth of an inch (3 mm) in diameter.

The first of the high uprushes, however, may be sufficient only to wet the sand to a depth of about half an inch, making a relatively impervious layer with air trapped beneath. When the next uprush comes and its water sinks downward, the air beneath the saturated zone is forced together in pockets as before. But now the sand surface is sealed, and instead of the air escaping as a series of small bubbles, it becomes one large bubble. Unable to escape as the pressure rises, the bubble lifts the sand above it into a low dome perhaps a quarter of an inch in height and 3 inches across or sometimes twice that large. One can tap these domes with a finger and feel them collapse, or slice them carefully with a pocketknife and see the dome structure.

Sand subjected to the action of moving water frequently forms parallel ridges and troughs, which are known as ripple marks. They are like small waves of sand and may be seen in sandy stream bottoms, on the face of a gently sloping beach when the tide is out, and beneath the surf zone. Ripple marks are commonly seen in ancient strata (sometimes with dinosaur tracks), and they have been photographed on the deep ocean floor in water 18,000 feet (5,500 m) deep. In short, whenever

Ancient tides were recorded in stone 3.2 billion years ago in the
Barberton Greenstone Belt, South Africa. These "herringbone" shapes
(cross sections of sand ripples above finger) were formed by alternating
tidal currents. The same dynamics shape beaches, build ripples, and
braid sediments today. *James Day*

there is sand underwater, ripples may be formed by the motion of the
water. They come in a great variety of shapes and sizes.

Ripple wave length may be as little as 2 inches (5 cm) and the height
a fraction of an inch, or they may form sand waves several feet deep and
50 or more feet (15+ m) long. Ordinarily, ripples are only a few inches
from crest to crest, and their size seems to be related to the size of the
sand from which they are formed, the velocity of the current, and the
amount of material in suspension.

Observation and carefully controlled experiments reveal that there
are two well-defined kinds of ripples:

1. *Current ripples*, which are formed by water flowing in
   a single direction, are asymmetrical with a long gentle

slope on the side from which the current comes and with a short steep slope on the lee side.

2. *Oscillation ripples*, created by the equal back-and-forth currents of flattened orbits at the bottom as oscillatory waves pass above, are symmetrical.

The ripple marks seen on beaches are almost always a complicated mixture of the two, because orbit size varies from wave to wave, and the direction of bottom currents is usually highly variable.

Since wave fronts generally parallel the shoreline, it might be expected that the underwater ripples would also parallel the shore, but this is true only part of the time. Even beneath the surf, the asymmetrical current ripples seem to be predominant. There, if the ripples are observed relative to a peg driven into the bottom, they are seen to be constantly in motion in an explicably random manner and direction. They are probably a prime mechanism in the movement of sand by water, but unfortunately ripples do not clearly indicate the direction of the main migration.

In fact, although the sand moves in the direction of the current, the ripple form does not. If the velocity of the moving water is more than about 2.2 feet (70 cm) per second, the vortex motion in the lee of the crest causes the ripple form to move against the current. When the velocity exceeds 2.5 feet (75 cm) per second, the ripples are swept away entirely as in an underwater sandstorm. All these minor beach features await you.

FAMOUS BEACHES OF THE WORLD

When one thinks of resort bathing, some famous beaches immediately come to mind: Miami Beach, Florida; Waikiki, Hawai'i; the French Riviera; Amalfi Coast, Italy; the Copacabana (Rio de Janeiro); Goa, India; Brighton, England; Coney Island, New York; Surfers Paradise, Australia; Maya Bay, Thailand; and Acapulco, Mexico. But what makes these beaches so desirable? Is it the kind of sand, the size of the waves, the temperature of the water, the climate, or the beauty of the surroundings? Or is it convenience to—or remoteness

from—population centers, the likelihood that other humans will be present, the low—or high—cost, or the reputation and advertising? In other words, are the main attractions natural or artificial, and which are the most important?

Those questions were asked by the government of Thailand, which felt its beaches were among the world's best and wanted to make this point with tourists and the outside world. To arrive rationally at some conclusion about what kind or color or size of sand is best, is subjective. People seem to have just as much fun on the shingle beaches of England or the rocky beaches of the Riviera as they do on the bright white coral sand of Bermuda, or the black-sand beaches of Tahiti. We are adaptable creatures and, within fairly wide limits, can have a good time in whatever way suits us at the moment. Some people are just as happy on the crowded beach at Coney Island at the end of a New York subway ride as others are on a remote island reachable only by private aircraft. The beauty and desirability of a beach are in the eye of the beholder. Under some circumstances you may like any or all of them; under other conditions—such as with a cold wind blowing or without your friends—you might not like any. For some, a beach in the Caribbean is less compelling than an ice-covered and wind-swept beach in Antarctica.

Waves of Plastics

Unfortunately, most beaches of the world are strewn with some form of plastic debris. Today, waves of plastic items are undulating at the sea surface, blown from afar by winds and quietly deposited on a beach face near you. A conservative estimate is that 8 million tons of plastic are dumped at sea per year. While not all plastic debris floats, on the remote Kamilo Beach on Hawai'i's Big Island, the top layer of sand is estimated be as much as 30 percent plastic by weight. It is estimated that more than 5 trillion visible particles float on our oceans. Some

RIPPLES Sand ripples are small waves of sand found wherever water flows over sand. Symmetrical ripples arise with the equal back-and-forth motions of waves, whereas asymmetrical ripples are created by water flowing in one direction. Great Exuma, Bahamas. *Jeremiah Watt*

Plastic debris is sorted with sediment, by size and density. The upper swash has diverse fragments intermixed with sand and tiny nurdles (plastic pellets used in manufacturing) all encroaching on the berm. Kamilo Point, Hawaiʻi. *M. Lamson/Hawaiʻi Wildlife Fund*

debris finds its way to the sea during the heavy rains that flush streams, culverts, and riverbanks of their refuse.

At sea, although forbidden by the United Nations Law of the Sea Convention, many vessels still dump their plastic garbage overboard. Some of the discarded refuse finds its way, transported by offshore winds and currents, to land and onto our beaches. Thousands of ships and shipping containers have been lost at sea, many containing plastics such as the pellets called nurdles, a raw material used in plastic manufacturing. Offshore, between Hawaiʻi and California looms the Great Pacific Garbage Patch, only one of five large garbage patches around the world, it comprises an estimated 250,000 tons of floating debris. Our famous beaches now sparkle with plastics.

Thankfully, there is some hope in reducing the world's plastic pollution. A Dutch company called The Ocean Cleanup has begun to collect plastics from the ocean's garbage patches and plastics at river mouths.

The Great Pacific Garbage Patch is formed by rotating ocean currents
that collect the masses of floating marine debris. Eastern North Pacific.
Caroline Power

It has deployed its river-interceptor systems in several countries in
Asia and anticipates many additional systems in other areas.

An additional promise is that the United Nations resolution was
signed in December 2017 by 193 countries as an effort to eliminate
plastic pollution in the sea. The nonbinding resolution is a step toward
a legally binding treaty to help reduce marine plastics and their devastating effects on the environment.

There is hope.

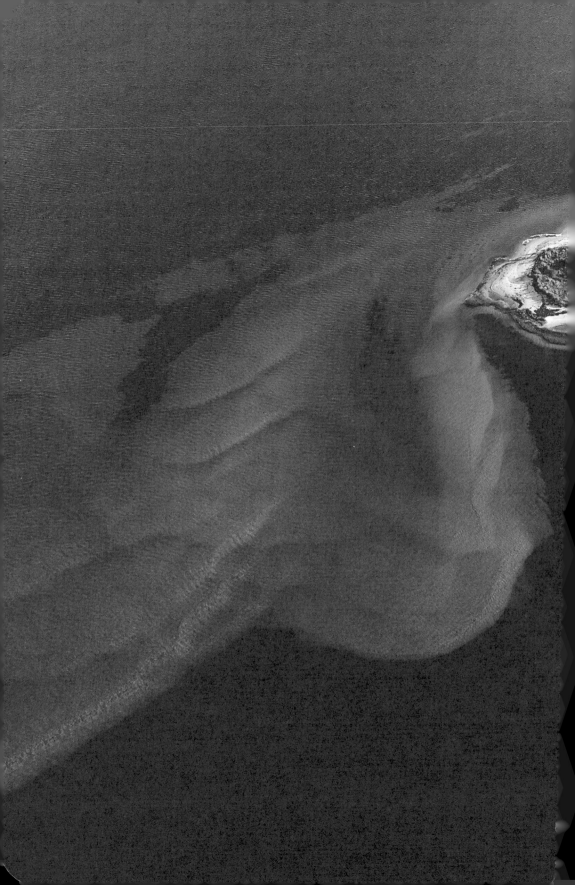

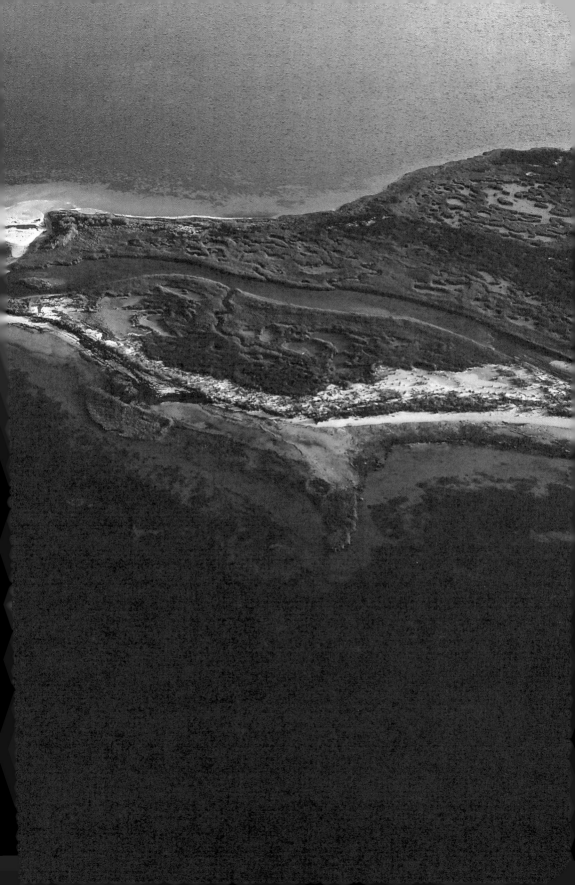

11 The Conveyor Belts of Sand

The movement of sand along a coast by wave-caused currents, called littoral transport, is responsible for most shoreline problems. Either sand is being removed from someplace that people wish it would stay or it is being deposited someplace where it is not wanted, or both. The processes are as old as the ocean and, by human standards, fairly slow; that is, years are usually required to make an appreciable change. As long as the property bordering the ocean is in the form of great ranches or public lands, no one pays much attention to a change in the position of the shoreline of 50 feet (15 m) over a period of fifty years. Its exact position is usually not known, and no great value is assigned to the land. When, however, a lighthouse or road is built near the water's edge so that there is a fixed object against which shoreline changes can be conveniently measured, the coastal dwellers first become dismayed by the loss of land, then alarmed.

When the large parcels are subdivided into small lots on the edge of the sea cliff and sold for high prices, as has happened in many places

PREVIOUS SPREAD: Coastal sediments move perpetually along the coast, driven by waves and currents, and deposited at the whim of nature. Goose Island, Chesapeake Bay, Maryland. *Bill Portlock/ Chesapeake Bay Foundation* around the world, the new owners are soon in an uproar when they find their land is disappearing at the rate of a foot a year. Although the processes of erosion have not changed, the rate of change has increased, as has the world's attitude toward coastal erosion. Even in the far north, the Bering and Chukchi Seas' barrier islands are losing the ice that once isolated them from erosion. Something must be done on a global scale, lest coastal dwellers, villages, towns, and cities accept further inundation and diminished land use.

The alarm of a rising sea level has been sounded in many places, including the Isle of Man in the Irish Sea. The small island's coastline is shifting and is losing land in some areas while it is being deposited in the south. The shifting is worrisome enough to create a National Strategy on Sea Defences, Flooding, and Coastal Erosion.

In England in the early 1900s, property owners whose land was being eroded by wave action clamored for the government to take preventive action. Their island was disappearing beneath the sea! They argued so loudly that a royal commission was appointed to study the matter. After making a careful survey, the commission reported that over a period of thirty-five years, England and Wales lost 4,692 acres (1,900 hectares) and gained 35,444 acres (14,350 hectares), giving a net gain of nearly 900 acres (360 hectares) a year. This finding seemed to prove that people whose land disappeared complained more loudly than those whose land was increasing. It must be admitted, however, that the land lost probably was good cliff-land on the open coast, which disappeared in a spectacular way, whereas the land gained was low, sandy, and not particularly valuable. Non-geologists are usually unaware that the very existence of a cliff is warning that erosional processes are at work, even though the changes seem to be very slow.

It is said that George Washington studied the erosion of the Long Island coast and ordered that the Montauk Point Lighthouse at the eastern tip be built at least 200 feet (60 m) back from the edge of the cliff so it would last 200 years. In the 1980s, the base of the lighthouse came as close as about 40 feet (12 m) to the edge of the cliff, but erosion-control efforts of the local citizens groups and government projects have delayed disaster.

Unfortunately, almost anything that either speeds up erosion of a coast or hinders the normal motion of sand alongshore affects all the other property within the same littoral zone. Any remedial action that does not consider the effects on the downstream beaches only causes more problems. Thus, the shoreline engineer, in addition to considering the complex, immediate problem of what action to take at any one place to keep the property owners there happy, must be careful that a solution does not create worse problems somewhere else. The best to hope for is a good solution for a few generations; eventually, the long-term geological process will overwhelm human endeavors. And climate change will make it worse.

SHORELINE EROSION

There are several mechanisms by which the sea attacks a cliff and makes sand.

1. Hydraulic and pneumatic action in which the pressure of water moving at high velocity against the cliff forces water into cracks in the rock and compresses the air that is trapped (this compressed air will sometimes shove even large blocks of rock away from the cliff into the waves);
2. The impact of water laden with rock fragments, which act as cutting tools against the cliff;
3. The abrasion or rubbing together of the fragments in suspension;
4. Grinding of the blocks that fall against each other as the cliff is undercut; and
5. Corrosion or chemical weathering of salt water and oxygen in the zone just above sea level.

But exactly how does rock become sand?

A vivid picture of the working of the "sea mill," which grinds large boulders to fine sand, was given by J. W. Henwood in an account of the visit he made to a mine that extended out under the sea in southwest England: "When standing beneath the base of the cliff, and in that part of the mine where but 9 feet (3 m) of rock stood between us and the ocean, the heavy roll of the larger boulders, the ceaseless grinding of the pebbles, the fierce thundering of the billows, with the crackling and coiling as they rebounded, placed a tempest in its most appalling form too vividly before me to be ever forgotten. More than once doubting the protection of our rocky shield we retreated in affright; and it was only after repeated trials that we had confidence to pursue our investigations."

Few people are privileged to listen to the surf from below, but similar sounds are created by large waves breaking in pocket beaches on steep rocky coasts. Where cliffs rise vertically from the sea, there are often slot-like depressions carved by the waves and floored with cobble beaches. Watching from the cliff above, one sees a wave break violently in the slot with much hissing and roaring. The churning water lifts cobbles as though they were sand grains and carries them upward in a surge of green and white froth. When this happens, one hears muffled *clocks* as the cobbles strike against each other. Then, at the top of the uprush, water and cobbles crash against the base of the cliff and the wave reflects. Down goes the water again, dragging its load of cobbles, causing them to clatter against one another with a loud, crackling sound. The observer is readily impressed by the violence trapped in the pocket and finds no difficulty thereafter in understanding how beaches have been created from cliffs by the relentless impacting and grinding of waves.

As the rocks grind against each other and cliffs are undermined, and as the moving sand abrades and then moves on, the coast retreats. Over the great lengths of geologic time it may be worn back many, many miles. Even in the short length of historic time there are many examples of substantial changes.

Old maps of the Yorkshire coast of England show the locations of many towns that have long since been swept out of existence by the waves, their former sites now represented by sandbanks far out in the sea. In 1829, the famous geologist Charles Lyell reported a depth of 20 feet (6 m) in the harbor at Sheringham where only forty-eight years before there had been a cliff 50 feet (15 m) high with houses on it. Now the harbor, too, is gone. Near Cromer, also in England, the sea cliff has long been retreating at a rate of 19 feet (6 m) a year and at Covehithe

Waves undermine cliffs at high tide and then grind their collapsed remains into sand. *Al Mackinnon*

and Southwold the erosion cuts the shore back 10 to 15 feet a year. During the great storm surge of 1953 (previously mentioned in connection with the Dutch dike failure) a cliff 40 feet (12 m) high at Suffolk retreated 40 feet (12 m) in a single night. A lower cliff lost 90 feet (27 m) to the sea during the same night. Such extremely rapid erosion is the result of unusually violent waves brought to bear against unconsolidated materials by a water level too high for the beach to offer its usual protection. By contrast, careful examination of the hard rocks of the Cornish coast indicates that they probably have changed little over the past 10,000 years.

The erosion and retreat of a shoreline is not always a steady year-by-year wearing away of a cliff and the seasonal removal of sand. It is important to appreciate such factors as bedrock failures, block falls, and sea-cave collapses. These types of slope failures are commonly associated with heavy rains and major storms, which generate large waves.

Homes on the verge of destruction in Pacifica, California. Millions of dollars, large boulders, and concrete have been a futile attempt to protect the land from the rising sea. *Josh Edelson /AFP via Getty Images*

As was seen in the Caribbean hurricane season in 2017, our changing climate includes frequent, stronger storms with heavier rain, greater storm surges, and bigger waves. In Solana Beach and Encinitas, California, the worst erosion occurs where groundwater from springs, leaky pipes, and shrub watering have lubricated rocky joints and fault planes. Deep tree roots and human activities (grading, building, and climbing on the steep cliffs) also accelerate coastal retreat.

A city map of Encinitas made in 1883 and updated in 1976 shows that about six city blocks have been lost to erosion, and at the maximum point the sea has cut in 800 feet (240 m). Much of this retreat followed the great flood years of 1884, 1886, 1889, and 1891. The county tax records show that as the property on the cliff edge eroded away, it was duly decreased in value, while the assessment of property immediately behind (with a newly improved view of the sea) went up. The loss of land continues for many coastal cities. Some

Waves and ice have worn down this Antarctic beach. Its mineral-rich sediments will provide nutrients for the sea's organisms. Cape Shirreff, Antarctica. *Kim McCoy*

municipalities accept the losses and retreat. Others, like Pacifica, California, have spent millions of dollars in repeated futile battles against coastal erosion.

Erosion is not all bad news on a global scale; it can also be a good thing for some organisms. The erosion of the land produces a nutrient-rich source of minerals for marine organisms. Dissolved solids, sediments, and iron and its compounds play important roles in the primary production for the world's oceans. The churning of the ocean waves helps grind and distribute these important nutrients from the land. The Amazon River and its sediments can be observed hundreds of miles into the Atlantic. In polar regions, the grinding of icebergs and sea ice against the land produces nutrients for deep-sea organisms.

THE LONGSHORE TRANSPORT OF SAND

Most longshore currents are generated by waves striking the shoreline at an angle. Although wave fronts bend as they move across shallow

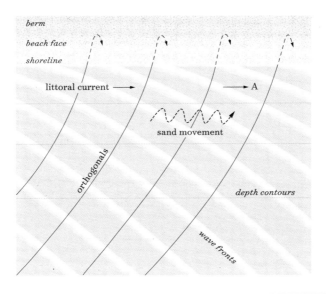

FIGURE 51: Waves approaching a straight shoreline at an angle are not completely refracted. The remaining alongshore component (A) is responsible for the littoral current. Paths of sand grains moving to the right with every wave are shown by dotted lines.

water and tend to become parallel to the shore, often the refraction process is not quite complete. When the wave finally breaks at a slight angle, either on a bar or on the beach, the water receives an impulse, part of which is in the alongshore direction. Therefore, the cumulative effect of many breaking waves is to move sand steadily alongshore (see figure 51).

Professor J. Munch-Petersen made the following analogy in the early 1900s: "One can get a good picture of the material movement if one looks upon the wave as an excavating machine and the wave current as a conveyor belt that moves the material the machine has loosened. Each wave machine lifts the sand and impels it in a more or less oblique direction, adding it to the conveyor."

Over the years Munch-Petersen modified the basic wave energy formula into one that he felt best described the ability of waves to transport material along a straight, sandy coast:

$$\text{Material moved} = \frac{KH^2L \cos \alpha}{8}$$

in which H is the wave height, L is the wave length, α "alpha" is the angle of wave attack, and K is a coefficient that depends on the size of material and the steepness of the beach.

Joseph Caldwell of the Coastal Engineering Research Center established a relationship between the amount of alongshore energy and the amount of sand moved, which gives similar results. It suggests that the energy expended in average weather to move the conveyor belt that extends from Point Conception to Los Angeles, California, is roughly 5 million foot-pounds per foot of beach per day (or for 100 miles, about 50,000 horsepower).

Along that part of the California coast, there is a delightful assortment of puzzling problems for the shoreline engineer, all created by an almost continuous littoral current from the west. This current is a product of the shape of the coast and the constant wave direction. North of Point Conception, the coast faces due west, directly into the wind and waves. But south of the point, the coast turns abruptly to the east so that the same winds and waves strike the shore at an angle. The current sweeps sand along the coast, and any structure that interrupts the flow acts like a dam, halting the flow of sand and causing the beaches to its west to grow. Beaches to the east of the structure are exposed to the inexorable waves and currents, and without new sand constantly arriving they retreat rapidly. A detailed study of sediment transport along the California coast from Point Conception to Point Mugu was published in 2009 with the support of the US Army Corps of Engineers and other government entities. It provides a comprehensive overview of sand sources, its movement, current management efforts, economic analysis, and future recommendations for the region. Let us look at two specific locations and examples.

The first major obstruction the moving sand encounters is the Santa Barbara breakwater. This location (Leadbetter Beach) has been the site of many research efforts, including the Nearshore Sediment Transport Study (NSTS) in 1980, a five-year project funded by the Sea Grant Programs of the National Oceanic and Atmospheric Administration, which helped develop nearshore instruments and measurement techniques, and meet national needs. The author's (McCoy) first job in oceanography was funded by NSTS. Many of the study's researchers and collaborators (Guza, Thornton, Dalrymple, Inman, Sternberg, Seymour, Dean) established the models for longshore transport still used in the twenty-first century.

As the sand moving along the beach from the west passes the tip of that structure, it abruptly encounters the deeper, quieter water that the breakwater was built to create. The wave-created turbulence that has held the sand in suspension ceases, and the particles are deposited just inside the end of the breakwater. But the filling in of the harbor is only half the problem. The beach to the east, deprived of its sand supply, is quickly stripped of sand and the soft cliffs behind are attacked. The solution for many years has been to dredge the sand from inside the breakwater and pump it into the first beach exposed to wave action, where it would remount the conveyor belt and continue on down the coast.

Extensive studies of the sand motion in the Santa Barbara area have been made. The principal effort was to attempt to correlate the rate of sand motion with wind and wave action. Wind and wave sensors were installed in Santa Barbara, and detailed beach profiles were made several miles upstream and downstream from the harbor. We soon discovered that the sand entering the harbor enlarged the perimeter of the sandspit in such a way that the growth was easy to measure. By making weekly surveys over a period of a year and plotting them as shown in figure 52 (page 290), we could then establish a relationship between the size of the waves and the growth of the spit. Because all the sand reaching Santa Barbara went into the spit, the growth was a direct measurement of sand transport along that coast.

On days when the breaker height was less than 2 feet (0.6 m) on the beach outside the breakwater, the spit grew at a rate of around 250 cubic yards (190 m³) of sand a day; if the breakers were over 4 feet (1.2 m) high, new sand moved in at 1,000 cubic yards (760 m³) a day; and during storm conditions the rate exceeded 2,500 cubic yards (1,900 m³) a day. The average sand flow during the year-and-a-half of our study was about 600 cubic yards (460 m³) a day, but data accumulated by the US Army Corps of Engineers over a period of years indicated long-term cyclic changes. Daily averages, obtained by dividing the amount of sand dredged out every two years by the number of intervening days, ranged from 400 to 900 cubic yards (300 m³ to 690 m³) per day. The highs

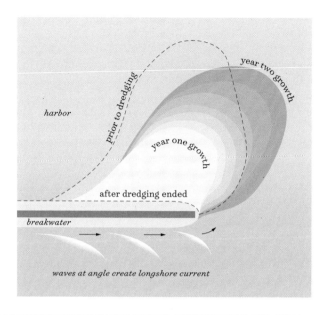

FIGURE 52: Growth of the sandspit inside the end of the breakwater at Santa Barbara, California.

and lows of these long-term variations appeared to come at about eleven-year intervals. Doubtless it is related in some way to the position of the weather system that controls winds and waves in the Pacific. This, in turn, is probably connected with solar activity of an eleven-year period, El Niño-Southern Oscillation, and other phenomena, but the mechanisms are not well understood.

Buried by the Berm

Leadbetter Beach in Santa Barbara, California, has been studied for decades. One winter, we were hoping to catch the first of the big storm waves from the Pacific. In anticipation, in the calm before the storm, during low tides, we deployed a vast array of sensors to measure sediments: cables, pipes, towers, communication, and recording hardware up and down the beach.

We waited a week. The storm arrived, and big, steep waves churned sediments. Surf zone currents moved the sand; tides rose and fell, and the berm was cut back. We recorded data and were elated with the results.

Elated, until some of the sensor data began to look strange. We waited for the low tide and assessed that many of the instruments were being buried under the new offshore bar formed by the large waves cutting back the berm. There were more high tides, yet bigger big waves cut away the remaining berm, and the offshore instruments and cables were covered by more sand. After about a month of deployment, it was time to call it quits and end the field season, but we were barred access. Our instruments were under thousands of tons of sand and irretrievable.

We packed up what we could and returned home. The berm had won, hundreds of thousands of dollars' worth of instruments were left in place. And we didn't retrieve our cache of gizmos until almost six months later after the storms had subsided, and the gentle, small waves had transformed the offshore bar back into a summer berm. Not all experiments end as planned. – KM

Beyond Santa Barbara, the moving sand encounters other obstacles, including Ventura, and Port Hueneme, where the deep-water port used by the Naval Base Ventura County is located. Here the breakwater spit is dredged out every other year by the Army Corps of Engineers, creating a great wave of sand to flow slowly down the coast; this causes some beaches to exist only on alternate years. Dredging began again in November 2019 to move 1.5 million cubic yards (1.2 million m³) of sand. The dredging will be repeated every two years—forever or until a storm of unprecedented fury destroys the breakwater.

The sand eventually reaches Santa Monica, much farther down the coast. That city also needed quiet water for a yacht harbor, but, aware of the sand problem and not wanting to become involved in never-ending dredging projects, it tried another solution. A breakwater was constructed parallel to the shore, several hundred yards out; the sand, it was hoped, would flow through the wide gap between breakwater and shore. It did not. In protected water, where the driving force of the waves ceased, the sand deposited. The result is that the beach behind the breakwater widens and itself becomes an obstruction to the movement of sand; downstream, the beach retreats. So, Santa Monica also uses a dredge to put the sand back into circulation and recently began dune-restoration efforts.

The general mechanisms of littoral drifting are apparent, so scientists are always seeking new ways of directly measuring the amount of sand suspended in the water and the depths at which it can be moved by various wave and current conditions. On the extreme Southern California coast, Douglas Inman of the Scripps Institution of Oceanography pioneered the study of the motion of sand in inshore waters. Spending much time on the bottom in diving gear, he observed sand being raised by passing waves and moved by currents; sampled the sand in suspension with various devices; and established the laws by which sand grains of various sizes respond to waves.

LITTORAL DRIFT

The coastal engineer is constantly confronted with variations on the problem of how to keep sand moving along a shoreline and at the same time prevent the shore from eroding. At any new shoreline facilities that stop the flow of sand, there will be trouble both at the place where the sand stops and the place where it would have gone. As with any problem posed by nature, the first step is to try to understand what is going on. If sand is filling a harbor, it is necessary to know where it comes from, how it is transported, and the rate at which it is arriving. If a beach is retreating, the engineer must know why before making plans to restore the sand. A study begins by obtaining answers to these critical questions. Then a plan of action must be developed and guided through the practical obstacle course that includes making legal, financial, and political arrangements, as well as the actual construction work.

Plans involving erosion-deposition problems inevitably rest on two solid pieces of knowledge:

1. Sand set in motion by wave-caused turbulence will settle out wherever a protective structure reduces wave action and turbulence ends.
2. If no action is taken on erosion problems, everyone shares the erosion. But as soon as one part of the shore is protected, the remainder of the shore must supply the sand.

Usually the engineer's first question is: What is the net littoral drift? How much sand will we have to deal with in this problem? The term net *littoral drift* refers to the difference between the volume of sand moving in one direction along a beach and that moving in the opposite direction (caused by shifts in the direction of attack of the waves). On a long, reasonably straight shoreline, the net drift is of primary importance, and on the Santa Barbara coast it is about 315,000 cubic yards (241,000 m³) a year.

Where the moving sand must cross an inlet, the total amount of sand in motion in both directions is important. Sand transport for Corson Inlet, an inlet through the barrier reef on the New Jersey coast, is a good example in volume per year: The southward-moving sand is 600,000 cubic yards (460,000 m³) and the northward-moving sand is 450,000 cubic yards (350,000 m³), resulting in over a million total cubic yards per year (765,000 m³/yr) movement of sand. The net movement is 150,000 cubic yards per year (115,000 m³/yr) to the south.

What happens? The sand shifting back and forth across the inlet is moved out of the littoral conveyor area by transverse tidal currents and is either carried out to sea, deposited as shoals in the channel, or is moved into the bay behind. In any case, the result is erosion of the beach, which must make up the deficiency by contributing the amount of sand withdrawn from the system. Moreover, the shift of the sand causes the inlet to migrate. If jetties were constructed to confine the flow of tidal water, at the end of a year there would be 150,000 cubic yards (115,000 m³) of sand accumulated north of the north jetty. This is the kind of situation in which it is worthwhile to think about installing a sand-transfer plant, with two objectives in mind: (1) to keep the entrance from shoaling, thus aiding navigation; and (2) to conserve 300,000 cubic yards (230,000 m³) of sand a year. Many such plants already exist in a number of places in the world.

There are places on the Southern California coast where such intermittent dredging has long been necessary. There was a need for enlarged harbor facilities in the Channel Islands area, so three human-made harbors have been built: Ventura, Channel Islands, and Port Hueneme harbors. Many of the beaches in the area are dependent upon bypass dredging. South of Port Hueneme, most sand is lost into the Mugu submarine canyon.

The Ventura Harbor has been carved from the dunes immediately upstream of Port Hueneme, and it too has parallel jetties, much like the first ones. But now the dredging situation is different, for a breakwater has been constructed parallel to the shore just west of the jetty; as at Santa Monica, this barrier causes the sand to deposit, but it also creates a zone of quiet water where a dredge can work without difficulty. Now, in one operation, the accumulated sand can be pumped around both harbor entrances and deposited on the beach beyond. This kind of sand-trapping, bypassing operation has come into general use in many parts of the world.

A similar but perhaps even more difficult problem is that the natural sand supply for many beaches has been cut off. South and west of the Los Angeles plain, the beaches had been nourished with sand brought down from the hills by small, intermittently flowing streams.

Ventura Harbor, California. Its jetties and offshore breakwater need periodic dredging to keep the harbor navigable for sport fishing and commercial vessels. The nearby Santa Clara River mouth complicates the sand transport in the area with large pulses of sediment following heavy rains. *Noble Consultants, Inc.*

But over the past century the local need for water and the demand for flood control caused these rivers to be dammed and their channels lined with concrete. (Later, at the end of this chapter, we will examine how dams and flood controls have created beach starvation in lieu of nourishment.) Today, the sand is trapped in reservoirs well back from the ocean, and the starving beaches have steadily retreated, with erosion extending as far as Newport Beach. The littoral current strips an estimated 200,000 cubic yards (153,000 m³) of sand a year from the beaches and carries it south, eventually dumping it into the Newport submarine canyon, from which it cannot be retrieved.

When the erosion first became serious in 1947, a million cubic yards (765,000 m³) of sand were dredged from the Anaheim Bay channel and deposited on the beaches to widen them. By 1963 the continuing attrition had made the situation on some of the beaches critical

again—seventy-five houses were smashed by waves in a single storm. So, another 3 million cubic yards (2.3 million m³) were dredged from Anaheim Bay and dumped on the upstream beaches. The littoral drift carried the sand along the shore toward Newport, a wave of sand that widened the beaches as it advanced. But human encroachment upon the rivers and coastlines now limits how much sand is available in the nearby bays and entrances. These seasonal sandy gifts to the beaches, the waves of sand from terrestrial sources, are only memories.

CURIOUS SAND FORMS

The configuration of a shoreline and the position of stream outlets are sometimes the result of wave refraction, especially along sandy coasts exposed to large waves. This happens because the height of the berm is

TOMBOLO This tombolo in Cantabria, Spain, was formed by the refraction of waves around the island. *Ana Tramont/iStock*

controlled by the height of the waves, and the berm acts as a dam. On straight, sandy coasts, streams will often flow directly toward the sea until they reach the beach. Then they will turn right or left abruptly and flow for a considerable distance—sometimes several miles—until the dam formed by the berm is lower. At that point the stream will cut through and enter the sea. If one looks into the reason why the berm is lower, it is usually found to be some variation in the offshore topography that caused the waves building the berm to be lower at that spot. Many beaches on the west coasts of Italy and the United States have streams that behave this way.

If there is a large rock or island a short distance offshore that protects the coast from wave attack, this has two important effects. First, it creates a zone of relatively calm water where sand or gravel will deposit. This material builds outward toward the island and finally creates a *tombolo* that ties the island to the mainland. A convex curve in the beach, such as forms behind offshore breakwaters or beached ships as sand builds outward, is called the tombolo effect. A second effect of an offshore island is it reduces the height of the berm in its lee so that streams in the area flow into the sea behind the island. For example, James Island on the Washington coast has a tombolo with the Quillayute River flowing out one side. This river used to adjust to wave conditions by flowing on alternate sides of the island until jetties were built to hold it in a permanent channel.

Sometimes the refraction caused by a submarine canyon will bend wave energy away from the beach opposite the head of the canyon, resulting in a low berm at that point. This happens at the Point Mugu lagoon in California. When the water in the lagoon rises, it cuts through the low berm opposite the head of Mugu canyon. However, there is substantial sand transport to the east in this area, caused by littoral currents, so the stream outlet then migrates eastward as sand builds up to the west and erodes to the east of the channel. In some years the entrance moves nearly 1 mile (1.6 km) eastward before a large winter storm creates a new berm, the lagoon begins to fill again, and the cycle repeats.

Another curious beach feature is the ridges that can form along river mouths. Most beach ridges (sometimes called *dune ridges*) are merely growth lines attesting to a series of abrupt advances by a shoreline—possibly single large storms or periods of exceptional littoral drift that brought in and deposited huge quantities of sand. These ridges are very common, and Cape Canaveral in east-central Florida is made largely of the ridges that have built up over the past few thousand years. Such ridges also exist at Clatsop near the south jetty of the river. The photo on page 299 shows the mouth of the Caravelas River in Brazil. These large beach ridge areas (each with multiple ridges) show three periods of distinct sedimentation. The ridge area on the left has an airfield on it.

Clatsop Spit, Oregon, has multiple ridges running parallel to the shoreline. Obviously, they were somehow built by the sea, but because they rise to 40 feet (12 m) above sea level it is evident that the sea used a special technique. The larger ridges are in pairs, and my examination showed elongated depressions on the seaward side of the large ones. What caused the ridges? One hypothesis is that the ridges were formed

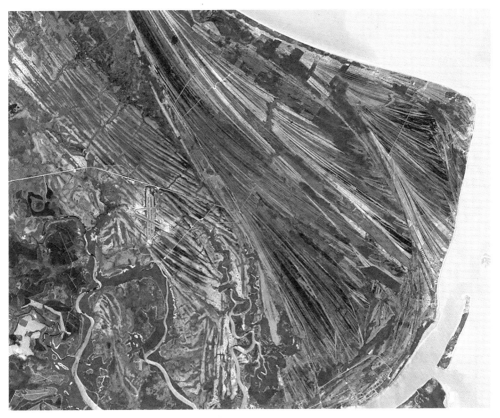

DUNE RIDGES Dune ridges record history in sediments as rivers
meander, winds blow, and sea-level changes. Caravelas Strandplain,
Bahia Province, Brazil. *NASA*

by ice shoved into the shore. Another more recent hypothesis is the
ridges were caused by an ancient tsunami inundation. A long time ago,
but within the past 4,000 years, during which the sea has been at its
present level, a great mass of ice or a tsunami probably transformed
this beach. Whatever the phenomenon, it repeated itself as each ridge
was formed.

TIDAL ENTRANCES AND BOTTOM FEATURES

Sand transport in tidal entrances is different from that on open beaches
where waves and wave-caused currents dominate. Typical tidal inlets on open coasts are in a state of dynamic equilibrium in which the

motion of sand in the entrance channel is constantly being adjusted in accordance with variations in the tidal currents.

In the 1970s, Steve Costa of the University of California at San Diego worked on the relationship between the velocity of the near-bottom water, the bed forms, and the amount of sand transport. In a series of model experiments conducted at the hydraulic laboratory of the Scripps Institution of Oceanography, he held the water depth constant and worked with sand of sizes ranging from 0.06 to 1.0 mm. He, and others before, have found that as water flows over a sand bed at increasing speed, there is a threshold velocity below which no sand is moved. As the velocity and viscous forces increase, individual grains begin to move short distances in a jerky fashion (*saltation*). A further increase causes asymmetrical current ripples to form, and a still greater increase brings about the formation of dunes, often with superimposed ripples. When the velocity is further increased, a transitional stage is reached; eventually the dunes are swept away and the bottom becomes flat. This produces sheet flow. At even higher velocities, long, stationary sand waves and antidunes are formed. The antidune has the curious characteristic that the dune form travels upstream while its sand grains are moving downstream (as they do in all the other bed forms). Many scientists have investigated and rechecked the above matters. The problem was that only in the transitional flat-bed situation could the rate of sand transport be simply related to the hydrodynamic regime.

In the other bed forms—especially ripples and dunes, which dominate tidal inlets—the sand transport mechanism depends on the turbulent flow of the water as it interacts with the bed forms. These features cause a large increase in the near-bed turbulence, which increases the average forces acting on the sand grains and creates fluctuating forces that further enhance the transport rate. The type of bed form present controls the way in which sand transport depends on current speed. Table 11.1 shows that sand transport is proportional to surprisingly high powers of water velocity. Actually, the absolute value of sand transport increases continuously with velocity.

In the dune regime, characteristic of most stable inlets, the transport of sand is very sensitive to small changes in velocity. A 1 percent change in maximum tidal currents results in a 6 percent change in transport rate. This remarkable amount of leverage caused John Isaacs to suggest that if a small additional flow could be added at a time that increased the velocity of the ebbing current, certain entrances could be kept free of excess sand. Tide gates or a pump system could add to the outflow of some bays to move sand at a lesser cost than periodic dredging.

TABLE 11.1
Power of V (Bottom Current Velocity) on Which Sand Transport Depends (after S. Costa)

Bed Form Regime	Sand Transport Proportional To
Ripples	V^{10}
Dunes	V^6
Translation and flat bed	V^3
Standing waves	V^6
Antidunes	V^6

Sand waves, dunes, and ripples on the bottom of Cook Inlet, Alaska, have been intensively studied. The features formed are similar to those found at tidal entrances, and their study is instructive. Cook Inlet is a very rough and rugged place. The average tidal range at Anchorage is 28 feet (9 m), and on the change of tides the flow of water in and out of the inlet reaches mean maximum velocities of 6 feet per second (2 m/sec) with 9 feet per second during tidal extremes. Winds of 100 knots and waves 20 feet (6 m) high occur, and in winter the surface is covered with floe ice. Beneath all this, the bottom currents create some remarkable bed forms: sand waves, dunes, ripples, and ribbons. Each of these types of bed-form fields covers only a relatively small area, perhaps a mile wide and a few miles long. Their boundaries are distinct, changing abruptly from one form to another.

Sand waves are the largest feature, having a wave length ranging from 30 to 3,000 feet (10 to 1,000 m) and a wave height between 1.6 and 33 feet (0.5 to 10 m). In profile these waves are strongly asymmetrical, with a steep lee or down-current side and a relatively gentle slope leading up to the crest. The crests are straight, or nearly so, and the elevations of troughs and crests are about the same. In other words, they are exceedingly regular features. The wave length to wave height ratio is

Underwater, seaward of the Golden Gate Bridge (San Francisco, California),
strong tidal currents and Pacific waves have organized centuries of
sediments into 30-foot (10-m) high sand waves (sonar image). *USGS* .

seldom more than 20:1. These bed forms are normally formed by moderate currents in sediments coarser than 100 to 200 microns (0.1–0.2 mm).
Sometimes these large waves have smaller ones on their flanks, which
may be parallel to the large crests or at a large angle to them, suggesting
variations in the current direction and speed. The waves seem to be similar to sand ridges reported from the North Sea, which are not believed
to have moved appreciably during the past three centuries.

The coarse sand is moved along the bottom when surface-current
velocities are over 6 feet per second (2 m/sec) and currents near the
bottom are nearly 1.3 feet per second (0.40 m/sec). This occurs only
during spring tides and perhaps winter storms. A significant increase
in transport power is observed with a decrease in water temperature.
The Cook Inlet study also found a similar but smaller sand form called a
dune or mega-ripple. Dunes have a wave length of 3 to 130 feet (1–40 m)

and height up to several feet. However, because they are steeper (their length-height ratio is less than 20:1) they show more apparent relief on the side-scan sonar images. The smallest features seen are ordinary current ripples, which seldom exceed 2 inches (5 cm) in height and 12 inches (30 cm) in length. Their shape depends on grain size and angularity and on the bottom current, which is often deflected or modified by the larger bed forms.

Sand ribbons were also seen. These are thin, narrow bodies of sand elongated in the direction of the current, commonly resting on a floor of coarser material. Sometimes there will be several such ribbons, roughly parallel, and sometimes sand in the ribbons will be in the form of small waves. Ribbons indicate that the sand supply is scarce and that the currents are more than about 3 feet per second (1 m/sec). Parallel ribbons are common in the North Sea and the Strait of Dover to depths of 500 feet (150 m). The main shifts of the deep-bottom materials come during unusually long-period waves (more than 14 seconds) or in the violent storms that occur every decade or so.

GROINS AND BEACH NOURISHMENT

For many years the accepted method of dealing with shoreline erosion problems was to build groins. A groin is a dam-like structure, usually a few feet high and about a hundred feet long, constructed perpendicular to the shoreline. Its objective is to retard the loss of a beach, widening it by trapping the passing sand. Groins can be made of timber, sheet-steel pilings, stone, or concrete. Some are built solid, to be impervious to sand flow; others—permeable groins—are constructed with openings that permit appreciable quantities of sand to flow through. Ordinarily, a system of groins is built to protect a long section of shoreline. Some parts of the New York and New Jersey coasts have groin fields extending for many miles. As material accumulates on the updrift side and the beach there widens, the supply of sand to the downdrift side is correspondingly reduced and the beach retreats. So, the solution is to build another groin, and another, and another. The slope of the beach face on the updrift side progressively steepens while that on the downdrift side flattens. Often the updrift side fills and overflows, the swashes of high

GROINS Groins influence wave action and interrupt the flow of sand, building it up on one side and carrying it away on the other. Long Island, New York. *John Wark/Airphoto*

tide carrying the sand over the top and spilling it on the low side, and soon a system of groins produces a series of short curving beaches that give the shoreline a cuspate appearance. As each groin fills, the sand bypasses the end and proceeds down the coast.

Although properly engineered groins can capture and retain sand, their effect is usually local and temporary. People with beachfront property in imminent danger of being washed away are understandably eager to take fast action, and without investigation they might build a groin in the hope of restoring their beach. The groin might be built in the wrong place and actually accelerate the erosion. The motion of coastal sand is more complex than one might suppose.

There are several ways to nourish beaches. The famous Waikiki Beach has been rebuilt several times, usually with sand trucked down from the north end of Oʻahu some miles away. Then, it slowly grinds itself into fine particles that drift along offshore, so that a fine-sand dump is formed at the seaward end of the old Ala Wai Canal. The wide beach at Long Beach, California, is made of sand dredged from the bottom, inside the distant breakwater; and Redondo Beach was similarly widened with sand scooped from Santa Monica Bay. Several beaches in San Diego County were nourished in 2001, including Torrey Pines State Beach, Del Mar, Leucadia, Solana Beach, Carlsbad, Cardiff, Encinitas, and Oceanside for a total of more than 1.8 million cubic yards (1.4 million m³). In 2012, some of these beaches were nourished again with more than a million cubic yards (800,000 m³) of sediment. Like other solutions to the problem of defending against the sea, this one is temporary.

Because groins rarely give a satisfactory long-term solution, they are no longer the preferred means of maintaining a beach. In the long run they are usually more expensive and less effective than a beach-nourishment program. So, the fashion is to add more sand, supplying it to the headwaters of the littoral stream from inland dunes or from the bottoms of nearby lagoons.

Many other shoreline construction projects, costing hundreds of millions of dollars, have been planned for shores around the world. Beach erosion is a problem of increasing importance as coastal land developments continue to be permitted, dams continue to restrict river sediments, and global sea levels continue to rise. Glacial ice, precipitation, and wave climate are all changing. These processes affect how and if groins are to be built and whether beach nourishment should continue. Some actions have unfortunately created sand deficits where sand is needed. We have starved beaches and dunes, the protectors of our coastal dwellings and places of commerce.

DAMS, RIVERS, AND DELTAS

It is impossible to separate dams from their rivers or deltas from their sediments. Hence, this section will somewhat blend subjects. Let us look at dams, rivers, and deltas from the perspective of littoral sediments and beach nourishment. The influence of dams, rivers, and deltas reaches far beyond the coast and extends into our social behaviors, environmental conditions, and political futures.

With few exceptions, beach material originates on land. In some areas, marine shells and skeletons of marine organisms (clams, mussels, snails, corals, sea urchins, crabs, barnacles, etc.) do contribute to beach material, but human activities have brought about great changes, some of which are observable at the beach face. We have constructed dams, restrained river sediments, modified river deltas, and changed marine life. Our actions have transformed the balance of power—we have upset the ancient battle of the waves and beaches. Today, we have the unprecedented, and simultaneous, waves of rapid population growth and climate change.

Climate patterns establish the locations where atmospheric water vapor condenses, forming the rain that becomes our streams and rivers. The rivers erode the land by attacking and churning the riverbed sediments; eventually the sediments reach the sea, forming our beaches and deltas. These river sediments, the products of hydrological flows from land, are met by their next adversaries—the sea and tides. There, at the edge of the sea, ancient beach cliffs collapse and surrender their sediments to the unrelenting forces of the waves, currents, and tides. However, in the twenty-first century the dynamics are more complex than humans have ever experienced. We are inexperienced strangers in our own environment. For most of the past four millennia, sea-level changes have been dormant, the weather patterns have been dependable, and rivers provided a seasonal flow of sediments to our beaches. All these have changed.

Today, most of the world's major rivers have been dammed and tens of thousands of lesser rivers and streams have barriers to natural flow. River sediments frequently fall prey to gravel mining and now struggle to meet the sea. Globally, sediment flows have been reduced by more than 50 percent because of dams and other human activities. It is estimated that the Santa Maria River in California delivers 68 percent less sediment to the sea as compared to prior to dam construction. River-borne discharges are met by the power of the local nearshore waves. Sometimes a delicate truce is struck between the river and the tides, and they collude to form an estuary. An estuary has a reversing flow; salty water flows inland at high tide, and at low tide the estuary flows back into the sea. Usually, estuaries have high biological productivity. If the river sediment discharge is great compared to the wave power, deltas can grow. Conversely, when there is a decrease in sediment discharge, deltas will retreat and beaches will die.

The Pitfalls of Coastal Development

It is no secret why our coastal areas continue to be developed in the face of increased natural disasters. Why do we continue to ignore the obvious? One answer is that the sirens of coastal change are muted by the trumpets of the revenue to be made from coastal development.

Housing and infrastructure projects produce jobs, and the resulting increase in tax revenues is warmly embraced by city councils, mayors, governors, and state and federal representatives. Bonds are issued by municipalities, and billions of dollars are earned by banks, developers, infrastructure construction firms, insurance companies, and real estate agents. Citizens enjoy the new landscape until nature tries to reclaim its lost ground. Nature will continue to restore balance with heavy rains, hurricane winds, immense waves, storm surges, compromised aquifers, and rising sea levels.

This poorly placed infrastructure will be reclaimed by nature at the cost of the taxpayer, not the developer. We must change funding and permitting policies for where we construct our houses. We must restrict the depletion of aquifers and the building of dams, and manage urban runoff—or our coastal communities will perish. Some cities have contradictory management policies; permitting sand mining upstream while at the same time paying millions for beach-nourishment programs downstream.

There are solutions. Many coastal municipalities, states, and countries have pioneered successful paths that we can follow or adapt. We do not need to invent new technologies to use the plans and methods that already exist. We need to organize, legislate, and implement new local and national policies. – KM

DAMS

Think of a dam as an energy accumulator. Instead of the water spreading its work slowly over the entire length of a river, the water and its energy are harnessed and used elsewhere for other purposes. Dams are constructed for a variety of reasons including flood control, agricultural irrigation, hydropower, and residential water supplies. Almost all large dams produce hydroelectric power for our domestic and industrial needs. Nevertheless, all dams decrease the peak water flow, alter the distribution, and diminish the supply of sediments. Occasionally, some dams are opened to provide increased water flow. On such occasions, without the natural sediment load the downstream riverbeds and banks can be excessively eroded and undercut. Even fish spawning is affected by the change in the size and distribution of sediments. These problems associated with the building of dams have been documented in the Americas, Africa, Europe, and Asia. Any change, reduction, or excess, in bedload causes imbalances downstream—where we are—at the beach.

The Aswan High Dam in Egypt has affected sediment transport and beach nourishment. Full capacity was reached in 1970, and the annual flooding along the Nile diminished, as did the sediment load. The reduced sediment load starved the adjacent fields of nutrients and resulted in an increased use of fertilizers, which were mostly manufactured abroad, and the increase in foreign imports resulted in a weakening and eroding of Egypt's currency, which persists to this day. Further downstream, where the Nile meets the sea, shoreline erosion continues. The Nile Delta at Rosetta and Damietta is estimated to have retreated more than 2,500 feet (800 m) and 4,200 feet (1,300 m), respectively, in the past twenty-five years. The continuing Nile Delta subsidence and rising sea level will worsen saltwater intrusion, reduce the amount of cropland, and cause profound landform changes. Even the offshore sardine industry has been adversely affected.

The dam at Aswan is a profound example of how nature will respond to human activities. Unfortunately, the hydropower and petrochemical industries have failed to recognize the pattern of nature's message to us, so obvious yet deeply encoded in our environment. The Nile water is an existential issue for Egypt. Far upstream, on the Blue Nile, Ethiopia continues construction on the Grand Ethiopian Renaissance Dam, with little consideration for the downriver repercussions. When finished, it will be Africa's largest dam with downstream environmental and political fallout guaranteed far into the future.

The Three Gorges Dam on the Yangtze River, China, is the largest hydroelectric facility in the world. It has changed the river's dynamics by increasing siltation upstream and decreasing the flow of water, sediments, and organic material into the East China Sea. *TopPhoto/ Alamy Live News*

The Yangtze River in China is the longest river in Asia at 3,800 miles (6,300 km). The Yangtze's Three Gorges Dam (TGD) is the largest power-producing facility in the world. It produced more than 93 terawatt-hours of electrical power in 2016. Prior to the TGD, the Yangtze's annual sediment and nutrient discharge was an estimated 490 million tons. After the building of the TGD, sediments decreased to below 180 million tons and a severe drought in 2006 reduced the sediment load to 9 million tons. The extinction of the Yangtze river dolphin is believed to be related to the TGD project. The Aswan High Dam and the TGD are just two large dams—according to the World Commission on Dams, there are more than 45,000 large dams over 50 feet (15 m) in height in the world.

Rivers

Rivers discharge sediments into the sea, including immense amounts by the Amazon, Congo, Ganges-Brahmaputra, MacKenzie, Mekong,

Mississippi, Yangtze, and Yellow Rivers. Each of these river systems establishes a unique but complex interdependent coastal environment for many biological and geochemical processes. When discharged into the sea, the energy of winds, waves, tides, and currents conveys the sediments into our biodiverse world. The largest freshwater discharge is from the Amazon River, discharging 18 percent of all river water and a whopping 10 percent of all river sediments. The Amazon's rainy season results in massive floodwaters exponentially increasing the discharge of nutrient-rich sediments and fluid mud. The North Brazil Current then carries these discharges more than 600 miles (1,000 km) along the coast, northwest past Suriname, French Guiana, Guyana, Venezuela, and into the Caribbean. In 1990, the US National Science Foundation funded the Amazon Shelf SEDiment Study and unraveled the fate of more than a billion tons of river sediments that pulse into the Atlantic Ocean annually. Unfortunately, there is a dam-building frenzy under way across the Amazon Basin including the 11,000-megawatt Belo Monte hydroelectric dam—the results will be devastating if the 400+ additionally proposed dams are completed.

Sometimes what happens thousands of miles from the ocean, removed from tides and waves, helps us understand what will happen at the beach. In Africa, the Niger River travels more than 2,500 miles (4,000 km) from the mountains of Guinea to the sea in Nigeria. As the river meanders through Mali it encounters a flat plain and spreads out into the Inner Niger Delta. Here it smothers the sands of the Sahara Desert for a hundred miles (160 km). It was here that the ancient city of Timbuktu arose and became the hub for trans-Saharan caravans for centuries. Today, massive sand and gravel harvesting in Mali is adversely affecting bridge stability, fishing, and agriculture. Thousands of Mali "sand divers" are upsetting the Niger River's equilibrium, driven by construction to satisfy the building needs of an increasing population. Similar dynamics are visible at many river deltas around the world.

The Carmel River in Central California is an excellent example of longer-term dynamics between river sediment, dams, water usage, house construction, estuary, steep beaches, and an offshore submarine canyon. The San Clemente Dam was built across the upper Carmel River in 1921 and provided drinking water, reduced stormwater flow,

The Amazon River's brown, nutrient-rich sediments flow along the coast and far out into the Atlantic Ocean, increasing the oceanic primary productivity and influencing the oxygen production and carbon dioxide consumption by phytoplankton. Brazil. *NASA/MODIS*

and protected the 1,500 homes around the estuary from flooding. Unfortunately, the increase in living quality for humans was offset by a precipitous decline in sediment flow and steelhead trout population. Prior to the dam, an estimated 12,000–20,000 steelhead climbed the river per year. By 1994, only 91 fish made it past the dam, and by 1997 the Carmel River steelhead were listed as threatened with extinction under the Federal Endangered Species Act. By 2008, heavy silting had reduced the amount of water behind the dam to only 5 percent of its original capacity. What occurred is a model for the future. A public-private consortium of citizen groups, corporations, and governments initiated a plan to restore the Carmel River. The river was rerouted and the dam was removed by 2015. The project cost $84 million USD, but the Carmel River's sediment flow has been restored, renewed estuary dynamics established, and the river has steelhead again.

Another example of human-river dynamics is the collapse in August 2000 of the Gaoping Bridge across the Gaoping River, Taiwan, China. The poorly constructed bridge had been affected by legal (and illegal) gravel mining and stormwaters, and the delicate balance of sediments and water flow had been upset. The Gaoping submarine canyon is just 1 mile (1.6 km) offshore from the river mouth. In this regime, any excessive sediment flows do not typically nourish the beach or produce offshore bars—the sediment is lost forever into the deeper waters of the submarine canyon. The disaster resulted in a large remediation project. Just a few years later in 2009, river stormwater discharges destroyed several offshore submarine cables.

In March 2001, when the Hintze Ribeiro Bridge in northern Portugal collapsed, killing fifty-nine people. It was determined that the combination of deferred maintenance, increased levels of stormwaters, and two decades of illegal sand mining caused the collapse. The illegal sand miners were prosecuted.

The building of dams and the harvesting of sediments can provide short-term gains but are frequently offset by heavy losses to the environment and to taxpayers. Global weather patterns are changing. Drought in some areas (e.g., California 2011–2017) and record rainfall with flooding in others (e.g., Texas, Hurricane Harvey in 2017, and extensive floods in Asia) further complicate river discharges and sediment loads. Droughts and dams can cause beach erosion. Floods can cause massive riverbank erosion and provide additional sediment for offshore bar formation. However, if the waves are large, the additional sediment reaching the sea can be forever lost into deeper water.

But there is hope in some areas. In 2011, two dams, the Elwha and the Glines Canyon Dams in the US state of Washington, were dismantled. It was the world's largest dam-removal project. In the following five years, more than 20 million tons of freely flowing sediment expanded the Elwha Delta into the Strait of Juan de Fuca. Over 1,000 other dams have been removed across the United States during the

The Elwha Delta in Washington state was replenished with fresh sediment after the removal of two dams upstream. The coast has assumed a vital new dynamic. *John Gussman* past twenty years. One project is the removal of four dams across the Klamath River system, which flows from Oregon into California and discharges its sediments into the Pacific Ocean south of Crescent City, California. It is the largest dam-removal project currently underway in the United States and will enable more than 400 miles of river to reprise its natural roles, including sediment transport. The benefits of the Arase Dam removal in Japan (completed in 2018) are apparent. Fish, prawns, seaweed, and sediment loads are approaching their prior levels as the environment continues to improve.

Deltas

A delta is formed when the supply of sediment from the land outpaces the local effects of the tidal flows and wave energy. It is where the waves and tides almost cede ownership to the land. Tidal amplitude-driven flows (currents) push oxygen-rich waters upstream into the meandering and braided channels of a delta. Some river deltas have small tidal amplitudes like the Mississippi, 1.6 feet (0.5 m); Mackenzie, 0.6 feet (0.2 m); and Nile, 1.3 feet (0.4 m). Others, like the Ganges-Brahmaputra, 18 feet (5.6 m); Indus, 14 feet (4.2 m); and Yangtze, 12.5 feet (3.8 m) have large tidal amplitudes. Tides play an essential role in the health of many deltas. Delta ecosystems possess great biodiversity and elevated primary production. Any harmony, the balance of natural forces at equilibrium, is easily upset.

Roughly 500 million people live in the low-lying regions of the world's deltas. The large-scale sinking of more than thirty deltas has been established using two decades of satellite data. These measurements have been augmented by land-based measurements using GPS. The demise of the world's deltas is primarily due to human activities that affect sediment availability and the flow of river waters. In some deltas, water and hydrocarbon extraction have accelerated sinking rates. The reductions in sediments are immense: Yellow River (China) is at 90 percent and the Nile River (Africa) is at 98 percent. It cannot get any worse than for the Colorado River (Mexico-United States) at 100 percent; all of the Colorado River flow has been sucked dry. None of the river's water reaches the Gulf of California (Sea of Cortés)

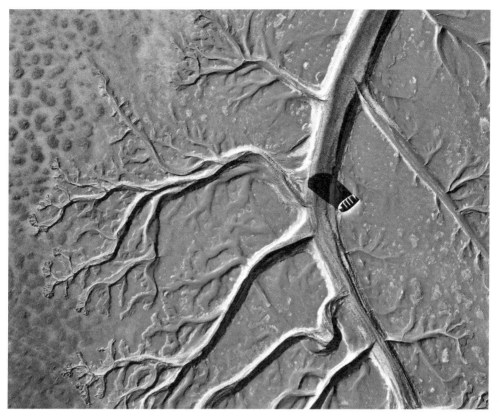

The once-lush Colorado Delta, now a dry riverbed of sand, was fished by the indigenous Cocopah tribe for a thousand years; only their abandoned fishing boat remains. Dams including Glen Canyon, Hoover, Davis, and Morelos (the final dam in Mexico) have diverted all the water for irrigation, growing populations, power generation, and industry. Sonoran Desert, Mexico. *Pete McBride*

in Mexico. Prior to the damming of the Colorado River in the early twentieth century, the northern end of the Sea of Cortés was a major wetlands ecosystem. The large tides (more than 20 feet) and strong winds (Chubasco) still churn the ancient sediments from past millennia.

The Mekong River Delta is the world's third-largest delta and home to 20 million people. The Mekong flows through China, Myanmar, Laos, Thailand, Cambodia, Vietnam, and into the South China Sea. It developed from estuary to delta over the past 5,000 years. Growth rates were between 52 feet (16 m) per year, up to 85 feet (26 m) per year in the southwest over the past 3,000 years. Today those same areas are in retreat. Satellite images of the Mekong Delta show that more than 50 percent of the delta is receding, retreating in some areas at a rate of more than 66 feet (20 m) per year. The supply of sediments, large-scale removal of mangroves, shrimp-farming activities, gravel and sand mining, groundwater extraction, and saltwater intrusion all threaten the Mekong Delta. As the flow of the Mekong diminishes, salt water is working deeper into the freshwater aquifers hastened by the rising tides. The Mekong is afflicted by dam construction projects as far upstream as China, yet the onslaught of cyclones will continue to attack the diminishing Mekong River Delta with large waves and surging storms.

Delta in Decline

In the summer of 2019, I drove from the Pacific coast over the Rocky Mountains, across the central United States until I reached Boston on the Atlantic coast. During that drive I passed through vast, open areas punctuated by cities, cultivated croplands, open mines, and underground fracking activities. I knew that all the effluent fluids from human activities (treated or untreated)—pesticides, chemicals, animal wastes, sewage, etc.—would either seep into the local aquifers or eventually drain into the great Mississippi River. From Boston I drove south along the Chesapeake Bay, North Carolina's beautiful Outer Banks, through Georgia and Florida, and eventually intersected the end of the chemical spigot: the Mississippi River and its delta.

Beyond Morgan City, Louisiana, where the delta reaches the Gulf of Mexico, the Mississippi releases its wastes into the Gulf. The Atchafalaya National Heritage Area sits upon an ancient river delta deposited more than 5,000 years ago. Today, the delta is under stress: levee construction has starved sediments and affected the water flow; erosion is a significant problem. A myriad of chemical pollutants including arsenic, benzene, mercury, and biological wastes intertwine with the lives of the residents.

Offshore is a large dead zone with low oxygen concentration, which suppresses aerobic life. The aging infrastructure, bridges, and river barges displayed signs for oil-field businesses such as Cameron, Schlumberger, Nautilus Pipeline Company, Neptune Gas Plant, Air Liquide, S.M. Energy, Knight Oil Tools, Munro's Safety Apparel, and law firms soliciting clients suffering from work accidents, chemical, and offshore injuries. These are the clear indicators of Mississippi Delta industry and commerce.

At the edge of the delta, at Burns Point Park, trees are dying and signs indicate no diving, no fishing, and swimming only at your own risk. There I encountered a woman in her early forties. I asked why nobody was swimming at the park. She replied, "I have been coming here since I was a child, but we do not swim anymore because of the toxic, flesh-eating bacteria. My daughter has never swum here, but it is quiet and peaceful; we like it." It is unclear why there has been an increase in flesh-eating bacteria cases worldwide, but some studies point to warming ocean waters. Today's Mississippi River delta is among the most-polluted deltas in the world. The US Clean Water Act of 1972 envisioned making all US waters "fishable and swimmable" by 1985.

We have failed. – KM

Even the world's largest and most populated delta, the Ganges-Brahmaputra Delta (which empties into the Bay of Bengal, Indian Ocean), with sediments almost 12 miles (20 km) deep, is in decline. Over 100 million people populate this delta. Some projections indicate that the combination of land subsidence and sea-level rise will displace 10 million people in Bangladesh by midcentury. As we saw in Chapter Four, storm surges in the Ganges-Brahmaputra Delta killed many hundreds of thousands of people in the twentieth century. The delta city of Kolkata in India, population of about 14 million, is under siege from rising waters.

Jakarta is the capital of Indonesia, and its largest city with 10 million inhabitants. More than 12 miles (20 km) of dikes have been built, but Jakarta is sinking so fast that the government now has plans to relocate the city to another location. Most of the area's river water is polluted (90 percent), resulting in the excessive pumping of groundwater and land subsidence. With the addition of global sea-level rise and heavy rains, Jakarta's future is increasingly wet. The Indonesian government has recognized Jakarta's suite of problems and is using funds to take action within a new legal framework. Jakarta is one of many large cities built on a delta facing decline.

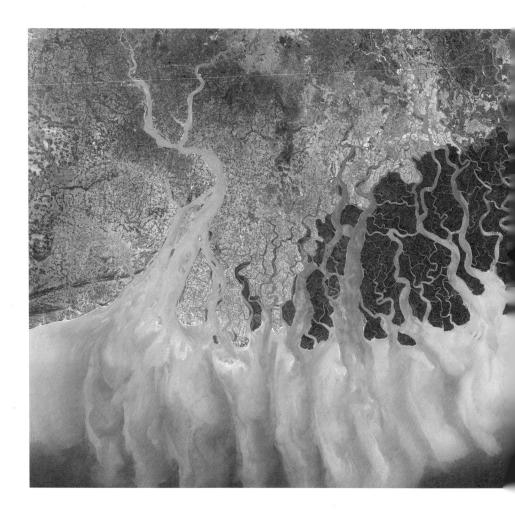

If we do nothing, what will be the history of deltas in the twenty-first century?

The continued damming of rivers, the mining of sediments, and the pumping of groundwater each provide the waves with an unfair advantage. We have swayed the balance of power and almost sealed the fate of the beaches, cliff faces, and deltas in favor of watery graves. Many older dams produce less electrical power and hold much less water than when originally built. Such dams starve the downstream coastal zones and deltas of the precious sediments that are needed to protect coastal

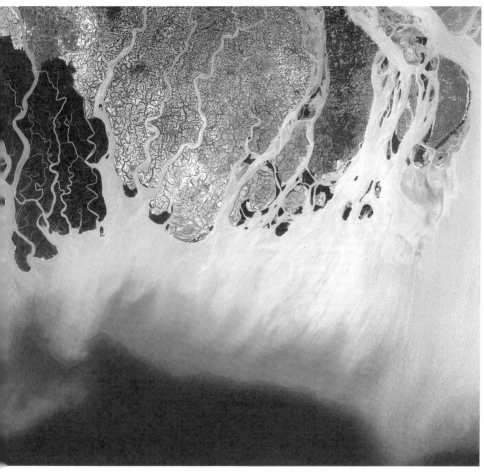

The braided Ganges-Brahmaputra Delta is the largest and most populated in the world. Storm surges and rising sea level have put its future in question. India and Bangladesh. *Universal Images Group North America LLC/Alamy Stock Photo*

resources from a rising sea level. In many areas, these naturally occurring sediments are the best and least costly solution to protect our coastlines. Continued negligence by officials will result in more coastal degradation. However, informed public engagement can elect new officials, overturn laws, and begin new practices for the public benefit. Gross Domestic Product is not a metric for environmental health or an indicator of an individual's happiness and well-being.

Age to age succeeds; Blowing a noise of
tongues and deeds, A dust of systems and
of creeds. – Alfred, Lord Tennyson

12   Man Against the Sea

The sea can be either friendly or hostile. It is calm and beautiful one day; furious and terrifying the next. On calm days one must not forget that before long, unleashed violence will follow. In this chapter we will consider what happens when waves smash against harbor defenses and shoreline installations, and what can be done to withstand the onslaught.

The solution to any problem begins with the attempt to understand what is going on. What is the nature of the forces? How do they act? What levels of energy are involved? We have gone into detail on the background information about the various kinds of waves, the way in which they refract as they enter shoal water, and the manner in which they are transformed into breakers. Now we must make use of this information in the design of coastal works that defend against the sea's attack. Experience is a good, although perhaps hard teacher, and it is well to begin by recalling some instances in which violent wave action has damaged shoreline structures in the past. These serve as a warning, reminding us of the extreme forces that waves may exert once in a decade or a century.

PREVIOUS SPREAD: Powerful storm waves have endangered structures near Land's End, in the United Kingdom, for centuries. Since the 1880s, one of the world's most important telecommunication cables has spanned the Atlantic starting from Sennen Cove (connecting London to New York). Today, undersea fiber-optic cables conduct terabytes of internet traffic. Incredibly, these cables have weathered shifting sediments, ships' anchors, and underwater earthquakes. *Barry Bateman/Alamy Stock Photo*

Then we will consider various means that can be used to defend our shores and harbors against the worst the sea is projected to do.

Waves Attack

Case histories of wave attack on man's coastal structures make for fascinating reading, for this aspect of the lore of the sea makes its great power most apparent. Many of the following examples were collected by D. D. Gaillard and presented in *Wave Action in Relation to Engineering Structure*, originally published more than 120 years ago in 1894. Lighthouses, by the very concept, are natural recipients of violent wave action because they often are built on rocky headlands or submarine ledges to keep ships at a safe distance. Some of them are called wave-swept towers, and it is understandable that lighthouse keepers should be an endless source of stories about fabulous waves and marine disasters.

For example: During the construction of the Dhu Heartach Lighthouse (*Dubh Artach* in Scottish Gaelic) in 1872, fourteen stones weighing 10 tons each, which had been fixed into the tower by interconnections and cement at the level of 37 feet (11 m) above high water, were torn out and carried into deep water. Windows in the Dunnet Head light station in north Scotland, 300 feet (91 m) above the water, are sometimes broken by rocks flung up by the waves.

South of Trinidad Head, California, a lighthouse is set on a rocky promontory 195 feet (60 m) above mean sea level, which doubtless seemed to its designers like a good, safe height. This illusion was shattered in 1913 when the lightkeeper reported: "At 4:40 pm, I observed a wave of unusual height. When it struck the bluff, the jar was heavy. The lens immediately stopped revolving. The sea shot up the face of the bluff and over it, until solid sea seemed to me to be on a level with where I stood by the lantern."

Several lighthouses are famous for having been swept away entirely by great storms and having been replaced more than once. These include the Eddystone light, Bishop Rock light, and the original Minot's Ledge light, off the Massachusetts coast, which was destroyed several times during construction and in 1851 crumpled into the sea carrying its two lightkeepers with it. The story about the light on Tillamook

Some massive structures do not withstand the forces of storm waves.
These rectangular concrete blocks were lifted and tossed into the
fishing port of Cudillero, Spain, in a storm in early 2014. *Jose Luis
Cereijido/EPA/Shutterstock*

Rock, a few miles south of the Columbia River mouth, is retold often.
The rock itself, several miles at sea, has nearly vertical walls rising to
a ragged surface about 90 feet (27 m) above the water on which the
lighthouse was built. During every severe storm, the entire rock shudders and fragments torn from the base of the cliff are thrown on top of
the rock. In a December storm a rock weighing 135 pounds (61 kg) was
thrown higher than the light, which is 139 feet (42 m) above the sea,
and in falling back broke a hole 20 square feet (2 m²) in the roof of the
lightkeeper's house, practically wrecking the interior of the building.
On another occasion, a fragment of rock weighing about half a ton was
rolled across the platform at the base of the building, 91 feet (28 m)
above low water, smashing a wrought-iron fence. In 1902, a keeper reported that water was thrown to a height of fully 200 feet (61 m) above
the level of the sea, "descending in apparently solid water on the roof."
Ten years later another keeper, investigating trouble with the foghorn

95 feet (29 m) above the water found it filled with small rocks. After the glass of the lantern was broken on several occasions by rocks, a heavy steel grating was installed 135 feet (41 m) above the sea, just below the lens, to prevent further damage. The Tillamook Rock light was deactivated in 1958.

As a result of this process of trial and error, which now extends more than 2,000 years, increasingly heavy construction has been used, and the problem of maintaining lighthouses on exposed rocks seems to have been reasonably well solved—except in such special cases as the Scotch Cap light described in Chapter Seven.

The only stories about wave violence that can top the lighthouse accounts are those about breakwaters. At Cherbourg, France, the break-water was built as an immense embankment of loose stone protected in places by 700 cubic foot (20 m³) concrete blocks. A 20-foot (6-m) high wall surmounts the stone embankment. During a severe storm many years ago, stones weighing 7,000 pounds (3,200 kg) were thrown over the top of the wall, and many of the concrete blocks moved as much as 60 feet (20 m).

In the Shetland Islands, a block of stone weighing 5.5 tons was detached from its bed situated 72 feet (22 m) above the high tide level and moved more than 20 feet (6 m). Another block weighing 8 tons was torn up and driven by the waves over several ledges with vertical faces 2 to 7 feet (0.7–2 m) high for a distance of 70 feet (21 m) at an average level of 20 feet (6 m) above high water.

In November 1950, extraordinary waves from a storm on Lake Michigan moved a concrete cap on the US Steel Company's breakwater at Gary, Indiana. This cap, 200 feet (61 m) long and weighing 2,600 tons, moved 3 to 4 feet (1 m) laterally when pounded by waves. Knowing the mass of the concrete, the engineers computed that the wave force required to move the cap must have been at least 1,680 tons of nearly instantaneous wave pressure.

Man Defends

The words that Captain Gaillard wrote more than a hundred years ago to describe the effects of waves on structures are as valid as ever:

"If wave motion is arrested by any interposing barrier, a part, at least, of the energy of the wave will be exerted against the barrier itself, and unless the latter is strong enough to resist the successive attacks of the waves, its destruction will ensue." No other force of equal intensity so severely tries every part of the structure against which it is exerted, and so unerringly detects each weak place or faulty detail of construction. The reason for this is found in the diversity of ways in which the wave force may be exerted and transmitted; for example, the force may be a static pressure, (1) due to the head of a column of water, (2) from the kinetic effect of rapidly moving particles of the fluid, (3) from the impact of a body floating upon the surface of the water and hurled by the wave against the structure, or (4) from the rapid subsidence of the mass of water thrown against a structure, which may produce a partial vacuum and cause sudden pressures to be exerted from within. These effects may be transmitted through joints or cracks in the structure itself, (1) by hydraulic pressure, (2) pneumatic pressure, (3) by a combination of the two, or (4) by the shock produced by the impact of the waves, which may be transmitted as vibrations through the materials of which the structure is composed.

In order to design any structure that will stand against wave action, one must have numbers that describe the amounts of energy involved and the magnitude of the forces imposed.

The energy in a wave is equally divided between potential energy and kinetic energy. The potential energy, resulting from the elevation or depression of the water surface, advances with the waveform; the kinetic energy is a summation of the motion of the particles in the wave train and advances with the group velocity (in shallow water this is equal to the wave velocity).

The amount of energy in a wave (per unit length of wave front) is the product of the wave length (L) and the square of the wave height (H), as follows:

$$E = \frac{\rho L H^2}{8}$$ where ρ is the mass per unit volume of water.

As the wave height doubles, the wave energy quadruples. A 12-foot (3.5-m) wave packs 295 foot-tons (9,000 kg-m) of energy. A big difference! For practical purposes the deep-water formula also applies to shallow-water waves.

A breaking wave can exert ten to fifty times higher loads as the same size deep-water wave, yet in most cases lasts only a fraction of a second. Today, we are armed with modern pressure transducers, electronics, and data-acquisition systems. The amount of data has increased and the results are frequently examined using statistics, yet the immense forces of nature remain.

THE DESIGN OF SHORELINE STRUCTURES

There are four major kinds of shoreline structures: jetties, breakwaters, seawalls, and dikes. All are made usually of some combination of rock and concrete. Jetties, usually in pairs, extend into the ocean at river entrances or baymouths to confine the flow of water to a narrow zone. If concentrated between a pair of jetties, the ebb and flow of tidal water keeps the sand in motion and prevents shoaling in the channel. A breakwater is a structure that protects a shore area, harbor, or anchorage from wave action. Often it is built well out from shore to provide a substantial area of quiet water. A seawall is built at the shoreline, separating land from water. It is the human-made equivalent of a rocky cliff, designed to protect softer material from erosion. A dike is a special form of impermeable breakwater that acts as a dam. When the zone it protects is pumped dry, it becomes a special kind of seawall. In the latter half of the twentieth century, the Netherlands addressed North Sea flooding with an immense moveable dike structure, the Rotterdam Barrier, which will be discussed later.

What can the engineer do to ensure that such structures survive against wave attacks? How many years of rising sea level will the structure withstand? There may be several possibilities, which, when combined, may provide the most efficient and optimal solution. The structure can be located so that its position, shape, and orientation give it the most favorable chance for survival. It can be built in such a fashion that wave energy is reflected or absorbed, or built with large and dense materials simply resisting the wave forces with its weight. The final plan is not necessarily the best possible structure to resist waves but

JETTIES A pair of jetties protect a small harbor from waves and unwanted sand movement. Green Harbor, Massachusetts. *US Army Corps of Engineers*

the best thing that can be done considering cost, time, use of existing facilities, kind of rock that is available, probable increase in the use of the installation, and, of course, political influences of many kinds. The engineer must fight hard for the best solution, especially in today's changing wave climate. Initial designs are frequently overruled, and the final structure may have substantial design compromises.

Here we will review only the technical areas within which the designer can maneuver to obtain the best engineering solution. No matter what other factors may ultimately influence the decision on what is to be constructed, this is always the first step.

The location of the structure comes first. For breakwaters, the underwater topography and the refraction conditions usually influence the design most. In Chapter Four, the process of making statistical studies of waves and drawing refraction diagrams was described. Briefly, the idea

BREAKWATER A breakwater's primary purpose is to protect a coast or harbor from waves. Lake Michigan, Chicago, Illinois. *Derrick Alderman/Alamy Stock Photo*

is to obtain as much information as possible about the waves arriving at the proposed harbor by hindcasting from old weather charts. Nautical charts will provide the initial information and establish the dominant wave direction and period, as well as the probable height of the largest or most damaging waves. Then, wave refraction diagrams are prepared; these are basic design data for the breakwater builder. Now taking into consideration the underwater topography, the general configuration of the coast, and the proposed locations of piers and wharves, the engineer can lay out various breakwater locations. If possible, the structure should not be exactly parallel to the wave fronts from the most probable direction. If it were, a wave could impact along the whole length of the structure at the same instant. Usually the stress will be spread out by presenting an angular front to the worst seas. In some cases, model tests will be made: first in a ripple tank, then in a larger wave tank.

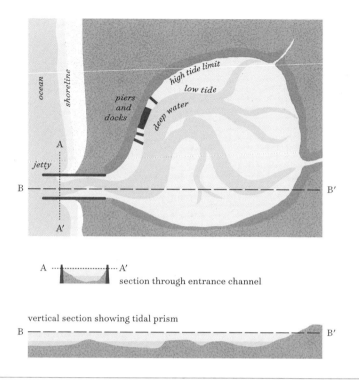

FIGURE 53: A protected harbor showing its tidal prism (the volume of water between higher high water and mean lower low water). The space between jetties, depth of water, and tides all influence sediment movement.

In locating jetty pairs, which often are parallel rocky structures extending directly into the ocean, a major problem is to space them properly so that the speed of the moving tidal water will be sufficiently high to scour the sand from the entrance channel and keep it deep enough for ship traffic. If the distance between them is too great, the water will flow at lower speed, and its energy will be diffused so that the channels will be shallow and sandbars will obstruct the entrance. On the other hand, if the jetties are too close together, the currents are likely to scour the sand from beneath the stones and undermine them. One can get an idea of the proper spacing by studying the natural channels. In the late 1930s, M. P. O'Brien examined a number of bays along the US Pacific coast and determined that a constant ratio exists between the area of the entrance section and the volume of the *tidal prism*, which is the volume of water inside a bay or harbor that

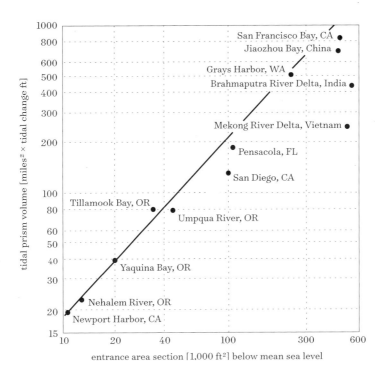

FIGURE 54: The relationship between tidal prism and entrance section for selected harbors, bays, and deltas. The line represents the ideal relationship between entrance area and tidal prism volume to avoid unwanted dredging.

is enclosed by planes of mean higher high water and mean lower low water (see figure 53). In other words, it is the average volume of water that flows in and out during a tidal cycle. Shoreline engineers who disregard this fundamental design law and arbitrarily space jetties at a greater distance than indicated by the line in figure 54 can look forward to years of dredging to keep open a channel that could be maintained by natural forces.

The next problem of the design engineer is to devise a shape for the structure that will reflect, absorb, or otherwise cushion the effects of large waves. It is possible to get rid of some of the wave energy by reflecting it; this approach may or may not be helpful, depending on the circumstances. If the structure in question rises almost vertically from water too deep for the waves to break, the wave can be reflected back to sea with little loss of energy. But few breakwaters are designed to

SEAWALL When properly constructed, a seawall can last for centuries. Alghero, Sardinia, Italy. *Diana Robinson/Getty Images*

reflect more than a small part of the energy. More often, they absorb the wave with rough faces of rock. Seawalls, on the other hand, usually are designed to reflect a substantial part of the wave energy that reaches them. Often, they have a series of steps, each about 2 feet (0.6 m) high, from which the water reflects without loading the whole structure simultaneously. Others have an overhanging recurved surface, which has the effect of throwing the landward-rushing water seaward, back upon itself. But if the face of the seawall curves so as to guide the water straight upward, as some old designs did, the water will fall back on the wall with great force, possibly damaging it and eroding the land it is intended to protect. A properly designed wave reflector has the effect of starting a new wave moving seaward, which tends to cancel the next oncoming wave it encounters, or at least reduce its force.

The resistance of rocks in a shoreline structure such as a breakwater or jetty to overturning or sliding is obviously of great interest to the shore-protection engineer. Dense rock (that is, having a greater weight

Tetrapods are like four-legged concrete sea monsters. The legs
interlock, making them hard to move by wave action, and they help
disperse wave energy on breakwater facings. Porto da Baleeira, Sagres,
Portugal. *Manfred Gottschalk/Alamy Stock Photo*

per unit volume) is far more useful, as the following example shows.
Compare a block of granite (density about 2.7) with a block of sandstone (density about 2.5), both with the same dimensions. The granite
weighs more than the sandstone, yet underwater, because of buoyancy
the effective weight of all objects decreases underwater. Because of
buoyancy, granite's effective density underwater is 1.7 but sandstone's is
only 1.5, making it less suitable as a breakwater material. Therefore, the
largest, densest rocks make the best breakwater material. Moreover,
rocks placed or connected in a line tend to support each other and can
resist more pressure than isolated blocks.

For some reason rectangular artificial blocks are nearly always less
stable than randomly shaped natural rocks of the same unit weight used
under the same conditions. This curious fact, long observed and eventually confirmed by systematic experiments, constituted a real challenge
to engineers. The question to be answered was: What is the best shape
for an artificial protecting block? Pierre Danel and others at the Neyrpic

Hydraulic Laboratory in Grenoble, France, began by setting down the properties that the new blocks (of whatever shape) should have. They decided these should be permeable so that the water could freely flow through. This solution would avoid creating any internal or back pressure, and it would reduce overtopping and reflection. The block should be shaped so as to have few plane surfaces; in fact, it should be as rough as possible to dissipate wave energy in turbulence; it should have maximum resistance. Combinations of blocks should interlock so they could mutually support each other.

In 1950, after many tests, a sort of sea monster with four tentacles was patented under the name of tetrapod. The most suitable proportions between legs and body were worked out with due consideration for the problem of manufacture, and not long afterward, tetrapods began to appear in shoreline structures. They are most useful when placed in a double outer layer facing the worst wave action over a core of natural stone and rubble. Today, tetrapods are used on the shores of many countries around the world. After the Fukushima tsunami-related disaster, Japan protected much of its shorelines with one-half- to 80-ton concrete shapes.

From the outside, to the uninitiated, the seawall may look like a rather ordinary piece of concrete with steps built for the convenience of bathers. Or the breakwater may look like a carelessly dumped pile of rock instead of carefully built-up layers. Figures 55 (opposite) and 56 (page 336), show the physical foundations and inner cores of several kinds of structures, but they do not show the true foundation—the years of hard-won experience and the elaborate computations that went into the design. In some cases, artificial reefs have been made by dumping rocks and construction rubble, or by sinking ships. It is a complicated subject and has many variations depending on local conditions and requirements, and this chapter obviously can do no more than give a hasty survey. If any reader has been stimulated to look further, an internet search will produce many of the proceedings from conferences on coastal engineering.

Another form of coastal defense is the sand dune. Like the beach itself, dunes can be nourished and encouraged by human acts. A zone of dunes behind a beach can act as a dike to prevent flooding of inhabited

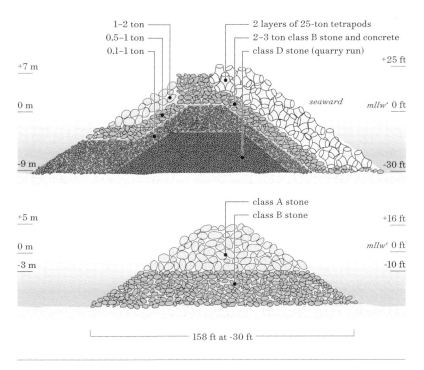

1–2 ton

0.5–1 ton

0.1–1 ton

2 layers of 25-ton tetrapods

2–3 ton class B stone and concrete

class D stone (quarry run)

+7 m   +25 ft

0 m   *seaward*   *mllw'* 0 ft

-9 m   -30 ft

class A stone

class B stone

+5 m   +16 ft

0 m   *mllw'* 0 ft

-3 m   -10 ft

158 ft at -30 ft

FIGURE 55: Sections through breakwaters. Top: The tetrapod-faced breakwater at Crescent City, California. Bottom: The rubble-mound breakwater at Morro Bay, California.

areas by unusually high tides or waves and, of course, the sand will resupply the beach if wave erosion becomes intense. Dunes are a sand savings account, held in reserve for the beach. An example of coastal dune erosion is at Fort Ord, California. Episodic collapses, due to storm waves and high tides, erode these dunes up to 6 feet (2 m) per year. With each collapse of the 65-foot (20-m) high dunes, the beach receives a nourishing gift of sand from above. When not acting in these protective capacities, dunes make fine recreational areas.

Dunes usually develop when the wind blows dry beach sand into the vegetation on the back beach. As the wind speed is slowed by the plants, the sand deposits around them. Thus dune vegetation is important and should be protected from dune buggies and careless visitors. In many dune areas, special plantings have been made to hold or increase the dunes. Alternatively, slatted fences, like snow fences, are used for the same purpose.

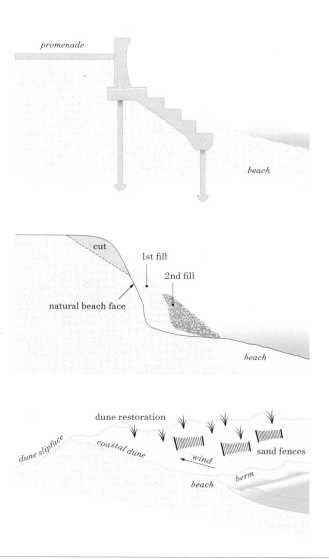

FIGURE 56: Three kinds of shore protection. Top: A concrete-step seawall with a re-entrant curved section, resting on sheet piling driven deep into the beach. Middle: A soft cliff is graded to a new slope and protected by a rubble-mound wall. Bottom: Coastal dunes can be restored by proper placement of sand fences and vegetation—nature will do the rest.

Some cities, such as Miami Beach, Florida, are built on barrier islands composed of porous stone covered by sand. Miami is about 4 feet (1.3 m) above sea level and is currently fighting rising waters and storm surges with pumps and seawalls. The city anticipates spending up to

COASTAL DUNES Coastal sand dunes, when left undisturbed by vehicles and construction, build naturally and protect the coast. Holkham Bay, United Kingdom. *Robert Harding/Nat Geo Image Collection*

$500 million USD more during the next five years. Its battle will continue for the foreseeable future.

The Humboldt Bay Jetties

The Pacific is the largest of the world's oceans, so large that long-period waves from as far as Asia and Antarctica may reach North America. The northwest coast of the United States from Cape Mendocino to the Strait of Juan de Fuca faces directly into the great waves generated by winter storms in the North Pacific. Many jetties and breakwaters have been constructed at the numerous bay and river entrances during the past two centuries. These concentrate the tidal flow in a single channel and protect ships from breakers in the dangerous transitional zone between the rough ocean and the quiet inland water. Rampaging breakers quickly reduce some of these structures to rubble, but a few have been persistently rebuilt by the US Army Corps of

The Humboldt Bay jetty faces into the open Pacific. Each T-shaped form (dolos) weighs 42 tons. The original jetty contract in 1888 commenced the first of many construction and repair periods, which continue to this day. California. *US Army Corps of Engineers*

Engineers, whose increasingly larger design wave is now a breaker 40 feet (12 m) high.

The history of the jetties at the entrance to Humboldt Bay at Eureka, California, 225 miles (360 km) north of San Francisco, is instructive. It is an example of the difficulty of maintaining structures where the sea prefers not to have them.

The beginning was not auspicious. In 1878, the Board of Engineers for the Pacific Coast arrived off the entrance aboard a survey ship to make the first official inspection of the proposed jetty site. Even though it was summer and the weather appeared to be moderate, the pilot refused to take the ship across the bar into the bay. He may have been influenced by an earlier report that said, in part, "It has been reported by masters of vessels that no such heavy seas have been encountered elsewhere in the world unless perhaps at Cape Horn. Waves have been seen to break in 8 to 10 fathoms of water." (This meant 40- to 50-foot breakers, or 12 to 15 m.)

The engineers decided to proceed anyway; the initial contract for the construction of a south jetty was signed in 1888, and within two years a north jetty was started. Both were to extend out to the 18-foot (5.5 m) contour, with their tops at the plane of high water. By 1891, the south jetty was 4,000 feet (1,220 m) long, and the north jetty was 1,500 feet (450 m). The latter used 28,000 cubic yards (2,600 m³) of brush and 100,000 tons of stone. Why brush?

The brush was used to make what was referred to as "mattresses." According to an eyewitness report: "Upon a grillage of poles, bound at each intersection, layers of bundles of brush are placed. The brush is about 12 feet (3.5 m) long, the bundles in each layer being at right angles with those of the next." The brush is then covered by rock.

Such was the way of jetty construction more than 150 years ago. The result was a ship channel 700 feet (210 m) wide and 25 feet (7.5 m) deep. Sixteen years later the channel had shoaled and the outer ends of the jetties were buried in the sand.

Between 1911 and 1915, the south jetty was rebuilt. Pilings could not be driven through the old stonework below, so the engineers shifted to the cap method. A concrete cap 22 feet (6.7 m) wide and 18 inches (45 cm) thick covered the jetties, and the south jetty ended in a 1,000-ton monolith. These were reinforced with railroad rails and secured with cables. When the job was finished, the jetty reached into 31 feet (9.5 m) of water, and the upper surface was 12 feet (3.6 m) above high water. More railroad rails were added, and a monolith of concrete 7 feet (2.1 m) thick and 30 feet (9.1 m) square was poured.

When the work was finally completed in 1927, the engineers must have stepped back proudly surveying their work and thinking, *There, that will do it!* But by 1932, repairs were needed; cubes of concrete, each weighing 100 tons, were cast and slid into place. By the 1940s, 12-ton tetrahedrons were used for further repairs. Then, during the winter of 1957–58, severe storms deteriorated both jetties, and repair became a major construction project. This time, 12-ton stones were used to raise the jetties to about 15 feet (4.6 m) above high-tide level. They were covered with a layer of 100-ton blocks, but most of these were washed away during the winter storms of 1964–65. By 1970, another major rehabilitation job was required. This time, extensive model studies were used and the adopted design consisted of placing 246 dolos armor units (similar to tetrapods) of 42 tons each on a slope of 1:5 against the heads of the jetties. Two layers of dolos units were used, with those in the upper layer carefully positioned to provide the greatest interlocking stability.

By 1978, there had been enough breakage and settling that the jetty was repaired, and yet again in 1983, 1988, and 1995. Such periodic additions, inspections, and maintenance will continue into the future. To date, more than 5,000 dolosse (plural for dolos) have become part of the Humboldt Bay jetties. The work continues—for the fiscal year of 2019 an additional $17.6 million USD was requested for the dredging and repair of the north Humboldt jetty. The work at Humboldt Bay faces rising sea level and battles with the stronger storms that climate change will bring. It is a Sisyphean task.

The ocean is huge, powerful, and eternal. Puny humans can scarcely expect to win by overwhelming it, and anyone who counters its attack with brute-force solutions is doomed to expensive disappointment. Rather, the engineer must try to understand how the sea acts and learn to take advantage of the geographic and oceanographic conditions. Then, on a battleground with nature, humans may be able to hold their own. The first and most valuable lesson one can learn about the sea is to not underestimate it.

THE DUTCH MAESLANTKERING STORM-SURGE BARRIER

In the Netherlands, the Dutch have created one of the largest human-made moving structures on the planet. The storm-surge barrier, the Maeslantkering, was built to protect Rotterdam and the surrounding areas from flooding. It is actually just part of a much larger Delta

The Dutch Maeslantkering is normally open; it can be raised vertically by flotation and then rotated closed (as shown here), thereby protecting the interior from flooding by North Sea storm surges. Rotterdam, Netherlands. *Frans Lemmens/Alamy*

Works project, a long-term visionary project to protect the Netherlands from myriad aquatic perils. The acceptable risks have been defined and codified into Dutch laws, making the defense against the sea a legal requirement.

The construction took six years, and the barrier was opened in 1997 at a cost of approximately $500 million USD (450 million euros). To close the barrier, the two large gates, each 790 feet (240 m) long and weighing almost 7,000 tons, are moved (floated) into place. When closed, the storm surge is blocked from entering the 1,100-foot (350-m) wide shipping channel. The Maeslantkering opening is wider than the passenger ship *Queen Elizabeth II* is long. The structures are a complex yet effective combination of long-term vision, funding, mechanical engineering, fluid dynamics, and computer software, but above all, a national commitment with great public support.

Its first real use against a storm surge from the North Sea was in November of 2007, when it spared millions in damage and countless disruptions to commerce, and it protected the population from ruin. Although it is expected to close because of a real storm surge only about every ten years, it is tested by closing it once a year. The Dutch Rijkswaterstaat authority closed all five of its storm barriers for the first time during a severe storm in January 2019. It worked as planned, saving billions in damages. Similar to the Dutch, the Russians have constructed the 16-mile (25-km) long Saint Petersburg Flood Prevention Facility Complex to protect the city against flooding from the Baltic Sea.

One must consider how to change the societal choices that continue to be made in favor of short-term solutions to long-term problems. The US National Flood Insurance Program should not masquerade as, or be a substitute for, sound building practices. The damages resulting from Hurricane Katrina in the United States in 2005 were estimated to have been about $100 billion USD. The reader may wonder why the financial and intellectual strengths of the United States failed to have the foresight to create and maintain adequate levees and pumping facilities for New Orleans. In these times of changing climate and sea-level rise, perhaps the Dutch are already upon the path that the world must travel.

The wind and wave forces exerted on offshore structures are immense; some are truly gargantuan. There is an unseen array of structures offshore for hydrocarbon production, windfarms, cables, and pipelines each of which must meet design criteria to interact with waves and sediments. Here are a few.

Hydrocarbons (gas and oil)

The *Troll A* natural gas production platform in Norway is 1,549 feet (472 m) high and weighs 1.2 million tons with ballast. It is the largest object ever to be moved by humans. Its design has withstood the North Sea's forces since it was towed 120 miles (200 km) to its worksite in 1996. Since then there has been a steady increase in the types, operational depths, and numbers of offshore structures. Roughly 30 percent of global hydrocarbon production is offshore. Thirty percent equaled 27 million barrels-of-oil equivalent (BOE) per day in 2015. The largest offshore oil-producing countries are Saudi Arabia, Brazil, Mexico, Norway, and the United States (in order of production).

In the United States, the second-largest gas and oil production area is offshore in the Gulf of Mexico. Hundreds of offshore structures and pipelines facilitate the extraction of approximately 2 million BOE per day. The pipelines crisscross the seabed and when pipelines fail, substantial environmental problems arise. On land, government "fracking" research has helped increase US hydrocarbon production by 10 percent in one year (2018 to 2019). In an interesting twist, some of the resulting glut of hydrocarbons will be used to increase plastic production (polyethylenes and others). The continued extraction of gas, oil, sand, and water for fracking slurries, land subsidence, coastal erosion, and plastic pollution will all prolong an environmental quagmire.

Windfarms

Many windfarms have been constructed offshore, each with typically 100 or more windmills (wind turbines), with some reaching more than 330 feet (100 m) in height. The detailed knowledge of the local wave climate, tides, sediment types, and sediment transport is essential for

survival of each windmill. Excessive currents can lead to the scouring of sediments, tower undermining, and failure. The constantly shifting sands and gravel affect the submarine cable integrity along its path to the land-based electrical grid. Offshore windfarms are commercially competitive, yet every one is at the mercy of the waves. Commercial systems are under way in Massachusetts, New York, Australia, China, Denmark, New Zealand, Scotland, and Sweden, to mention a few. The largest energy company in Denmark is Ørsted A/S, and it is also the largest windfarm operator in the world. Some of its windfarms have giga-watt capacities, and Ørsted A/S anticipates having 4.1 terawatt-hours of annual production from its Hornsea 1 project by 2020. Small countries can have a large impact on how the world harvests its energy.

Cables and Pipelines

Thousands of miles of subsea cables and pipelines lie beneath the surface and interact with the seabed. Each cable or pipeline has an entry and exit point into the water, some are trenched and buried in sediment, yet all are affected by coastal dynamics. The route and de-ployment must be carefully planned to minimize wave forces, sediment movements, and seabed scouring. There are more than 750,000 miles (1.2 million km) of fiber-optic subsea cables and more being planned. Companies such as Amazon, Facebook, Google, and Microsoft invest in subsea cable projects. If you have ever accessed the internet or made a telephone call using digital telephony, you have used a subsea cable. For decades, internet traffic via subsea cables has been of vital interest to intelligence services including the National Security Agency, Govern-ment Communications Headquarters, and others. Each cable's route has been planned with attention to coastal and deep-water dynamics yet almost one hundred cable failures occur each year.

Subsea pipelines are much larger in diameter than cables. Pipelines transport valuable hydrocarbon products (natural gas and oil) from off-shore and terrestrial production sites. The Nord Stream 1 pipeline runs

Offshore windfarms convert the wind's energy into electricity. These structures and their undersea cables must withstand wave forces and shifting sediments. Off Block Island, Rhode Island. *Dennis Schroeder/ National Renewable Energy Laboratory* from Russia across the seabed of the Baltic Sea to Germany. Nord Stream 1 is 760 miles (1,200 km) long, 4 feet (1.2 m) in diameter, and has been in operation since 2011. Nord Stream 2, although almost complete, is entangled in a geopolitical storm which surpasses the complexity of wave forces and shifting sediments. Longer than Nord Stream, the EastMed pipeline, is planned to be 1,080 miles (1,900 km) long, extending from offshore Israel to Cypress and into Greece. The EastMed pipeline is at the opposite end of the unpredictable geopolitical spectrum and such complexities are beyond the scope of this book. An internet search for maps of "subsea pipelines" will reveal interesting connections. Nevertheless, tens of thousands of miles of subsea pipelines form international webs of commerce. Subsea pipelines are complex in regulation, have many sediment-related failures, and exist unseen by most.

A clever use of seabed pipelines is for air-conditioning purposes. Cold deep waters are pumped to the surface and passed through heat exchangers (cold radiators) to satisfy office and hotel air conditioning demands. Several remote and urban areas including Hawai'i, Bora Bora, Cornell University (using lake water), the city of Toronto (using Lake Ontario), and Stockholm, Sweden, are using such systems. The effluent water is returned back to the ocean (or other body of water) at the thermo-neutral depth. The method is very efficient and system costs are repaid within three to seven years. This technique has great promise both environmentally and economically. The City of Honolulu, Hawai'i, has several projects to provide cooling for city and state office buildings.

Ship Motions, Stability, and Losses

Ships and seagoing structures are the largest moving objects ever created by humans. Immense planning, design, finance, labor, and energy (mining metallic ores, making and transporting steel, etc.) contribute to the existence of these giants of commerce. A brief discussion of ships—how they interact with waves and how they are lost at sea—can help understand and prevent environmental damage as a result of these losses. Regardless of country of origin, our coastlines and beaches will struggle to absorb the blunders of humans at sea.

Consider the motions of a ship in response to the movements of the sea surface. A ship can make six basic motions, three of translation (*surge*, *sway*, and *heave*) and three of rotation (*roll*, *pitch*, and *yaw*). Surge means the ship moves directly ahead or astern along the line of its keel; sway means it moves broadside as though being pushed sideways to a pier by a couple of tugboats; heave means it moves vertically up and down. If you push a toy boat in the bathtub directly downward and then release it, the excess buoyancy will cause it to heave upward and it will bounce up and down a few times at the water surface with its natural period of heave.

As for the rotational motions, roll means the deck tilts to one side and then the other; pitch means the deck tilts fore and aft with the bow down and then up. When a vessel yaws, it rotates about a vertical axis, as, for example, when it is overtaken by a wave not directly astern. Then, for a moment the stern will move sideways a bit faster than the bow—something like the rear wheels of an automobile skidding on wet pavement as the whole car rotates.

As the motion and the restoring force alternately move the ship, it oscillates with a periodic motion. The up-and-down oscillation in the heave of a toy boat has been mentioned. In that case the restoring force is explained by Archimedes' principle: a floating body displaces a weight of water equal to its own weight. In pushing the boat down, an excess of water was displaced. When that force heaves it up, the inertia carries it too far, but after a few oscillations it uses up the inertia in each direction and comes to rest. The periods of roll and of pitch are usually of more interest to the ship designer; they also involve excess displacement and inertial overrun.

As an experiment using a toy boat in the bathtub, or perhaps a dinghy alongside a pier, push down on one side and then let it freely rebound. It will rock back and forth several times; the time in seconds to complete each cycle is the period of roll. A similar experiment on the end (bow or stern) of the boat will give the natural period of pitch. With large ships these periods are discovered in a more complicated way, but the principles are the same. Every ship has its own natural periods that derive from its shape and the way its structural weight is distributed and its cargo is loaded. There are good reasons for wanting to know these natural periods; the stability and steadiness of the ship depend on them.

Stability is the tendency of a ship to return to its original upright position after it has been tilted by external forces. *Steadiness* is the tendency of a ship to minimize motion caused by the sea. These two factors conflict with each other, and nearly every ship design is a compromise between the two. That is, most people want a ship that is steady (moves so little they are not likely to get seasick) but at the same time they want to be quite sure it is stable (will not roll over). The word "stability" is frequently misused by persons meaning "steady."

The measure of (initial) stability of a ship is expressed as its *metacentric height*, commonly known as GM. The GM expresses the relationship between the center of gravity (G) and the vertical center of buoyancy—known as a ship's *metacenter* (M). A large GM results in a very stable ship; the disadvantage of this is a fast response to the sea, which causes unwanted stresses in the ship, the likelihood of cargo shifting, and uncomfortable passengers. Such ships are said to be stiff, such as barges and workboats (e.g., for crane operation, transferring equipment, etc.).

On the other hand, a small GM produces a tender ship, which is comfortable because it has a poor response to waves and a long, slow roll. The disadvantages are reduced safety if a compartment is flooded and a greater chance that large waves will break over the decks. Passenger liners typically have small GMs for comfort (with high decks and lots of watertight bulkheads).

Most supertankers and bulk cargo ships are long enough to bridge several wave crests and massive enough to have a slow response. Among the reasons for concern about the natural period of the ship is that this reduces the chance that the ship's period will be the same as that of the waves. A ship with a roll period of, say, twelve seconds would roll violently if it ran parallel to the crests of a group of twelve-second waves because each roll would be reinforced by the next wave (as when pushing a swing at just the right time). Ships frequently run at an angle to the waves, which makes the effective wave period longer. The ship's speed can make the effective period longer or shorter, depending on

PLIMSOLL MARK The secret language of ships. The Plimsoll
mark certifies safe load lines for ship or offshore structure operation.
The TF stands for tropical fresh; below that is an F for fresh
water, and an LR for the classification society: Lloyd's Register.
Nieuwland Photography/Shutterstock

whether the waves are coming toward the ship or moving with it. The
result is often that the bow slams and great vibrations or shudders run
through the ship. In Rudyard Kipling's words: "The ship goes wop with
a wiggle between."

Depending somewhat on the size of the ship, the designer may treat
it as a structurally rigid beam. That is, it will have enough strength so
that it would not break if it were supported at only the midpoint or only
at each end. All ships bend somewhat, because steel is very elastic.
With a single wave crest under the midpoint, the ends droop a bit; this
is called *hog*. If the bow and stern are supported by crests, then the
center section *sags*.

One ship has been built that sacrificed all other considerations to
decouple it from sea motions. This is *FLIP* (Floating Laboratory Instrument Platform, see pages 186–87) of the Scripps Institution of

Oceanography. It is shaped like a long pencil 350 feet (107 m) long and 12.5 feet (3.8 m) wide. *FLIP* has two positions of stability: horizontally on the water surface much like a ship (it is not self-propelled) or it can flip, by flooding some compartments, to a vertical position to avoid wave motion. When vertical, this oddly shaped platform is extremely stable in ocean waves. A slim structure placed vertically is also referred to as a "spar buoy."

A ship is less likely to survive a storm at sea if it is overloaded. Insurance companies and ship classification societies require the ships for which they are responsible to not be overloaded. Ship designs must withstand varying wave conditions, water densities, and buoyancy while at sea. Cold water is more dense (and hence provides more buoyancy) than warm water; as is the ocean's saltier water. There is an easy way for an inspector to determine proper loading, devised by Samuel Plimsoll in the 1800s, and amazingly it is still used today. It is a paint line identified by a circle on the side of every commercial ship that shows the waterline for allowable loading in various seas. This line and circle is known as the Plimsoll mark (see photo on page 349). Usually it is identified by the initials of the classification society (A and B, for example, denoting the American Bureau of Shipping) that has approved that position of the line. Alongside the Plimsoll mark there are other qualifying marks, all related to wave and water conditions.

Why Ships Sink

Ships are relatively safe when they are in deep water and well offshore. They are most likely to be lost by running aground (usually at night) or being driven ashore by high winds or waves. Consequently, when large ships are at anchor in small harbors and a big storm is forecast, they will often put out to sea, preferring to take their chances in open water to the possibility of dragging anchor and going aground.

Amazingly, collisions with other ships, wave-induced structural failures, and explosions are still frequent. Any change in a vessel's stability makes it more vulnerable to the waves and frequently precedes its sinking. Loss of stability accounts for many of the large ship, structure, and shipping container losses. More than 5 million containers are aboard ships at sea at any given moment. Each year 1,000+ shipping

The MV *Rena*, running at 17 knots with 1,300 containers aboard, altered course to avoid unfavorable tidal currents and ran aground on a New Zealand reef. Built in Germany for an Israeli company, later flagged in Malta and then Liberia, she was ultimately owned by a Greek firm and leased to an Italian company. Bay of Plenty, Tauranga, New Zealand. *Graeme Brown/Maritime NZ*

containers are lost at sea. Some have highly toxic contents and pollute the environment long after they have disappeared into the deep. The toxic items transported by sea include all types of chemicals and radio-active materials.

In October 2011, the MV *Rena* lost an estimated 900 containers after running aground off the coast of New Zealand. Perhaps the largest loss of containers was 200 miles (300 km) off the coast of Yemen in 2013. The MOL *Comfort* was under way between Singapore and Saudi Arabia, broke in two, and subsequently sank in more than 10,000 feet (3,300 m) of water. Nature attacked the MOL *Comfort* with furious winds and waves. The significant wave height was over 18 feet (5.5 m), and the wind was force 7 Beaufort (near gale at 30 knots); quite substantial for a ship of more than a thousand feet (316 m) traveling at 17 knots. The ship

Some shipwrecked containers do not sink; they drift ashore and disintegrate on the beach face. These came from the stranding of MSC *Napoli*, carrying 2,400 containers of which 150 held hazardous substances. Branscombe, United Kingdom. *Leon Neal/AFP via Getty Images*

was a total loss and cost the insurance company more than $300 million USD. More than 4,000 containers were claimed by the sea.

In January 2019, one of the world's biggest container ships (almost 200,000 tons), the MSC *Zoe*, was lucky, only losing 270 containers overboard in a North Sea storm off the coast of the Netherlands. It could have been much worse—there were many thousands of containers onboard. Some of the containers managed to wash ashore into the UNESCO Biosphere Reserve of the Wadden Sea. The locals managed to legally scavenge all sorts of fun cargo including televisions, car parts, shoes, and plastic toys. Unfortunately, three of the lost containers were transporting hazardous materials—"organic peroxides" are usually not considered compatible with a biosphere reserve.

Superships, including some of the largest ships ever built, have also disappeared at sea. For example, on December 30, 1975, the three-year-old combination ore-bulk-oil carrier MS *Berge Istra*, 224,000 tons, en route from Brazil to Japan, suffered a series of explosions east of the Philippines and sank in very deep water. Its fate might never have been known except that two crew members, picked up in a life raft nineteen days later, described how this huge ship—1,030 feet (314 m) long and 164 feet (50 m) in beam—sank in a few minutes. The inquiry tentatively ascribed the cause to "undetected gas pockets detonated by sparks from missiles or machinery." Just a few years later in October 1979 the sister ship, MS *Berge Vanga*, sank under very similar circumstances with the loss of all hands. Since these disastrous explosions, ships that carry combination ore-bulk-oil have become less popular.

But we are more concerned here with ships sunk by the violence of the sea and the increasing intensity of storms in the twenty-first century. Some ships have been lost because they were poorly designed or not built exactly as designed. In an effort to make ships go faster, they are made slimmer—which may also mean that they roll over easily. Design trade-offs are made that balance the available shipbuilding facilities, capital, maintenance costs, fuel efficiency, speed, and seaworthiness. When seaworthiness is compromised, things break.

Mishaps arise when the whole ship structure is so weak that the ship flexes as waves pass beneath it. With a wave crest at the bow (front) and the stern (back), the center of the vessel is relatively unsupported (or vice versa), and there is always minor bending of the ship's structure. But if the ship flexes too much it can come apart; the deck crumples or pulls in two, the ribs spread, the seams open, and its stays rip out the sheer strakes. The ship can quickly become a mass of scrap. The extreme wave forces rip ships apart as they lose buoyancy and are driven into the deep by gravity.

If the crew is not vigilant, a ship can easily be overwhelmed by the conditions at sea. Winds at whole-gale force, with velocities of 50 knots or more, can by brute force disable or sink a ship. Or a ship may be low in the water because of leaks or overloading. With low freeboard, the lee rail easily goes under; then perhaps the cargo will shift or the deck will leak, and the list will increase until the ship cannot right itself again. Or a ship blown hard before the wind, perhaps built with a sleek bow for speed, will drive into the back of a huge wave and be taken by the sea in one gulp. On other occasions, not necessarily in very high winds, a ship will suddenly pass through a weather front into stronger winds coming from a very different direction. Open hatches or scuttles, now on the low side, let the sea pour in and the ship is gone.

Lessons from History

During the wars between France and Britain (in the period between 1793 to 1815) the British Royal Navy lost more than 300 ships by mishap but fewer than a dozen by enemy action. This is at a time when the total of vessels in the navy at any one time was about 250; the figures suggest that the fleet was completely rebuilt every twenty years if the entire shipbuilding program only replaced ships lost to the sea.

Relying on Lloyd's Register Group Limited statistics—for merchant ships in the last days of sail, when ships were probably better designed and built—we have the following shocker: 10,000 sailing ships insured in England were lost in various parts of the world in the five years from 1864 to 1869, nearly a thousand without a trace! Of the 130 vessels that departed European ports for the Pacific in the summer of 1905, 53 had vanished in Cape Horn waters by the end of July.

An astonishing number of modern ships will disappear at sea every year. Even with careful inspections by insurance companies, globally available weather broadcasts, and voyages measured in a few days instead of weeks, some will be declared to be "missing without a trace." Some losses will remain unknown.

In October 2015, news headlines covered the loss of the container ship SS *El Faro*. The *El Faro* (length 791 feet, or 240 m) left Jacksonville, Florida, and headed to San Juan, Puerto Rico. It steamed directly into Hurricane Joaquin, took on water, listed to port, and sank despite having recently been certified by the American Bureau of Shipping. All thirty-three crew members perished along with several hundred containers, shipping trailers, and automobiles. Surprisingly, a precise number of seagoing shipping containers lost at sea each year remains unknown.

Lloyd's of London posts a ship as "missing" if it disappears without known cause, leaving no survivors and no substantial wreckage. "Posted as missing" is the formal death certificate of the crew and the clearance for the insurance claim to be paid. Today's ships are immense objects, now surpassing 1,300 feet (400 m) in length and weighing more than 200,000 tons. Each year billions of tons of cargo are moved around the world. According to the *Safety & Shipping Review* issued by the global insurance company Allianz, roughly 100 large ships are lost by sinking or foundering each year, with more than 1,000 lost over the past decade. Ships move like blood in veins, nourishing the world's commerce, all at the mercy of wind and wave.

All bodies of water experience storms. The disappearance and sudden loss of the 25,000-ton SS *Edmund Fitzgerald* in 1975 is immortalized in song and in numerous writings including *Fitzgerald's Storm* by Dr. Joe MacInnis. The SS *Edmund Fitzgerald* sailed the Great Lakes of North America and remains the largest ship to have sunk there. The Lake Superior storm that sank the ship had sustained winds of over 60 mph (100 km/hr), gusts over 80 mph (130 km/h) that created waves reported at 35 feet (11 m) in height. The wreck of the SS *Edmund Fitzgerald* was located and was determined to have sunk due to the intensity of the storm, cargo loading, and construction techniques. One of the largest vessels to be lost at sea with no trace was the British registered MV *Derbyshire* in September 1980 in Typhoon Orchid off the coast of Japan. It was just under 1,000 feet long (300 m) with a deadweight tonnage (DWT is the maximum allowable weight of cargo, fuel, ballast, etc.) of 160,000 tons. Efforts by the Seafarers Union and relatives of the victims located the MV *Derbyshire* fourteen years later in 13,000 feet (4,000 m) of water off the coast of Okinawa. It was determined that storm waves damaged piping, which caused the buckling of a large cargo hatch that led to massive structural failures.

In January 2018, the Panamanian-flagged vessel *Sanchi* carrying 136,000 tons of ultra-light oil collided with another vessel 160 nautical miles (300 km) off of Shanghai, China. Explosions followed, the *Sanchi* burned for a week and sank with a loss of all thirty-two crew members.

The above stories describe ships unexpectedly getting into trouble, mostly from storms and being overwhelmed by the sea; they take on water and sink. The ratio of foundering (with some survivors) to missings (with no survivors) is about 11:1, and not all losses are reported.

Possibly the greatest test of design, ingenuity, and endurance against waves at sea is ocean sailing. The Vendée Globe is a nonstop,

single-handed yacht race around the world. It begins and ends in Les Sables-d'Olonne, France, and circles the globe in the Southern Ocean. The current record is just over 74 days. Another true battle with the sea is The Ocean Race (formerly the Volvo Ocean Race and before that the Whitbread Round the World Race). The designs for these lightweight offshore sailing vessels include carbon fiber hulls, masts, and sails. They use specialized materials such as titanium, Dyneema®, Kevlar®, and stainless steel to resist the sea. The fury of the winds drives the boat forward, while below many tons of lead keel and complex hydrodynamics attempt to balance the forces of gravity, buoyancy, and viscosity.

The forces of wind and wave are massive: mast compression is more than 60 tons, windward shroud tension is more than 25 tons, and forestay tension can reach 17 tons. Peak speeds of over 30 knots (55 km/hr) are reached, sometimes surfing down waves in the dark and covering distances of more than 500 miles (800 km) per day, propelled just by wind and wave! The sailboats have a crew of only nine, and they race day and night. The sailors are constantly doused by waves, deprived of sleep, and stressed to their limits. The race includes periods of more than twenty days without landfall, including in the Southern Ocean. This southernmost of the world's oceans encircles Antarctica, resulting in an endless fetch, with terrifyingly huge waves with significant wave heights reaching over 33 feet (10 meters) and maximum heights of more than 70 feet (21 m). The grueling battle of sailing machine and the sea lasts more than nine months while enduring an ominous onslaught of more than a million waves. The boats transit 40,000 nautical miles (74,000 km) around the world—not all vessels or crew members survive.

Many coastal losses are caused because of tidal currents or navigational errors due to negligence. The infamous oil tanker *Exxon Valdez*, with a length of 987 feet (301 m), ran aground in 1989 due to such an error. Eleven million gallons (roughly 260,000 barrels or 35,000 metric tons) of oil spilled into Alaska's Prince William Sound, resulting in an ecological disaster compounded by tidal currents that frequently reach 2 mph (1 m/sec) or more.

The Vendée Globe sailboat race requires a superb understanding of winds, currents, wave forces, equipment design, and human endurance.
Kurt Arrigo/Rolex

Unfortunately, only a small number of ship losses make it into the mainstream media. However, the cruise ship *Costa Concordia* made international news in January 2012. The 952-foot (290-m) cruise ship struck submerged rocks in the Tyrrhenian Sea near the Mediterranean island of Giglio. More than 4,000 people were onboard and luckily only thirty-two perished. The captain, Francesco Schettino, was sentenced to sixteen years in prison for the events related to the sinking. The *Costa Concordia* was refloated in a very complex operation and subsequently cut up for scrap. The sinking incurred more than $1 billion USD in losses.

Any large structure can sink, even offshore deep-water oil-drilling rigs designed to be stable in the harshest wind and wave conditions. There are an estimated 3,000 offshore oil-drilling rigs around the world. According to the US Energy Information Agency, collectively they produce millions of barrels of oil per day—almost 30 percent of the world's oil production. Since the year 2000, deep-water production has more than tripled. Each time a rig sinks, it provides an oil slick on the surface of the water, and surprisingly some portion of the heavier oil sinks, sending an unobserved black death into the depths. The oil rig *Ocean Ranger* sank in a storm in the North Atlantic in 1982 with the loss of eighty-four lives. The storm had 55-foot (16.7-m) waves. The rig sank after being struck by a rogue wave estimated to have been 65 feet (20 m) high. Another rig, *Petrobras-36*, sank in 2001 off the coast of Brazil after an explosion followed by bad weather. When it sank, it was the largest semisubmersible rig in the world.

The explosion of the *Deepwater Horizon* in the Gulf of Mexico in April 2010 was one of the largest marine environmental disasters ever. It was also legally very complex and costly: The rig was owned by a Swiss company, built in South Korea, registered in the Marshall Islands, operating in US waters, and leased to a British company. *Deepwater Horizon* cost over $500 million USD to build, and the accident killed 11 of the 126 crew members. The US government estimated it caused between 12,000 and 19,000 barrels of oil to be spilled per day—for many months. Other estimates put the spill total at 4.9 million barrels (670,000 metric tons). The oil company BP (formerly British Petroleum) created the Deepwater Horizon Oil Spill Trust ($20 billion

Flawed operations on the *Deepwater Horizon* caused methane gas from below to explode in a billow of smoke and flames. Continued instabilities led to a massive oil spill and the sinking of the *Deepwater Horizon*, and the loss of eleven crew members. Gulf of Mexico, off Louisiana. *NOAA*

USD) to help mitigate the negative environmental effects across the US Gulf of Mexico. More than two years later, in September 2012, turbulence from Hurricane Isaac cast oily material from the *Deepwater Horizon* spill upon the land. Despite the massive cleanup efforts, the environmental damage to the Gulf of Mexico above and below the surface still persists.

Globally, more than 200 offshore oil-drilling rigs have had blowouts, collapses, fires, groundings, or sinkings due to structural problems and/or weather conditions. Most of these disasters are far enough offshore to remain unnoticed by the public, yet they have occurred in the Mediterranean, Persian Gulf, Atlantic, North Sea, Pacific, Indian, and Arctic Oceans and off the coasts of Angola, Borneo, India, Nigeria, Norway, Alaska, Brazil, and Mexico. No part of the world is left untouched; untold tons of oil spill into waterways, seas, and oceans each year—from the tropics to the Polar regions. These incidents pollute our environment and contribute to the changes in our global climate. Fortunately, the United Nations Convention on the Law of the Sea provides a legal framework, albeit with limited jurisdiction over these substantial offshore activities. The International Maritime Organization (IMO) is the "United Nations specialized agency with responsibility for the safety and security of shipping and the prevention of marine and atmospheric pollution by ships." The IMO has begun to monitor the carbon dioxide emissions from ships, and it estimates that shipping emits 3 percent of all anthropogenic carbon dioxide. The IMO has implemented measures to reduce the amount of allowable sulfur in ship fuel oil by 80 percent (starting in January 2020) to help reduce the sulfur dioxide emissions from ships.

Increased activism by individuals and governments is needed to diminish the ongoing damage to the deep-water and coastal environments. A shift away from hydrocarbon energy sources will reduce oil spills on the waves and abate the clumps of oil on the world's beaches. The immensity of this issue must not be ignored.

From the beginning of January 2013 to the end of December 2013, according to Lloyd's List Intelligence Casualty Statistics, almost 100 ships were lost. The British newspaper *The Guardian* provides another estimate for that same period quoting the "World Casualty Statistics" with 138 "total losses"—a ship beyond repair or recovery. John Thorogood, from IHS Maritime, estimates eighty-five of those sank, "in that the vessel actually went at least partially below the sea in a fairly traumatic manner." A rough estimate is that about two ships sink per week. In addition to these larger vessels, an unknown but larger number of small commercial and fishing vessels are lost at sea.

Even ships designed to sink will sink. There have been at least nine nuclear submarines that have sunk, exploded, or been scuttled (intentionally sunk) at sea. Their reactors, now silent, still leak radioactive material into the oceans.

Lloyd's Registry of Merchant Ship Losses is full of interesting reading about losses at sea. The IMO estimates there are about 85,000 vessels (over 100 gross tons) working at sea, and some of these vessels sink, with an annual loss of about 2,000 lives.

Life at sea is not easy. Anyone who has doubts about the types of troubles at sea need only read the "Perils" paragraph of the American Institute Hull Clauses, dated 2009, which is used in insuring vessels and structures at sea. The Hull Clauses includes all sort of dangers. Here are a few: "the Seas, Men-of-War, Fire, Lightning, Earthquake, Enemies, Pirates, Rovers, Assailing Thieves, Jettisons, Takings at Sea, Arrests, Restraints, and Detainments of all Kings, Princes and Peoples, Barratry of the Master and Mariners, Detriment or Damage of the Vessel..." and so forth.

New ways to be "lost at sea" have been added, including "Volcanic Eruption, Contact with Satellites, Aircraft, Helicopters or similar objects, or objects falling therefrom." Not only are mountainous waves, tempests, and typhoons to be feared, but now a celestial stream of helicopter and satellite parts!

Throughout history the winds and waves, and other natural forces, have sunk far more vessels and destroyed more coastal structures than armed conflicts or terrorist actions. But as we have seen, there is a new nature to Earth's winds and waves. Our climate has changed; our ports, offshore structures, commercial ships, and oil platforms are under siege. All will need to withstand larger, more violent storms and support higher operational costs. It is clear that some coastal metropolises will be uninhabitable as losses increase and nature's responses evolve. Coastal dynamics has a commanding influence on insurance rates, commerce, our environment, and politics. Now is the time to take action. Soon, human influences on the environment will be irreversible; with powerful dynamics beyond our control. We are twenty-first century hunter-gatherers and as before, we desire water, food, energy, and information. Sea-level rise, eroding coastlines, and compromised aquifers will force many to wander, driven by disrupted weather patterns and energy sources. Thankfully, we have an abundance of information which contains a sea of solutions.

Epilogue

Many of the most pleasant hours of the authors' lives have been spent in watching waves and examining beaches, trying to understand them. Walking and meditating, photographing, digging, and surveying—the curious quest of both authors spanning a period of over seventy years has been great fun in spite of uncounted saltwater drenchings and numerous scientific disappointments. They have covered thousands of miles of shoreline in fifty countries, on all continents without losing any of the fascination, and without producing any hope of complete understanding. The subject is too complex. But somehow there is satisfaction in being aware enough of the ways of waves and beaches to detect the special softness of a new layer of sand underfoot that means the berm is building or to observe a slight change in the appearance of the breakers and think, *There must be a new storm in the Gulf of Alaska.*

The inner peace that comes with the quiet contemplation of a beach on a still, calm morning or the feeling of exhilaration that comes from

PREVIOUS SPREAD: Bioluminescent plankton (dinoflagellates), when excited by the turbulent energy of breaking waves, emit light. Some of their light radiates back into the universe where it all started. Laguna Beach, California. *Stan Moniz* riding a great wave in a small boat is more reward than most ever know. Fortunately, the beaches of the world are swept each night by the tide. A fresh surface always awaits the student, and every wave is a masterpiece of originality.

It will ever be so. Go and see.

A note on the third edition: Willard Newell Bascom passed from this life on 20 September 2000. His ashes were spread into the surf at the merging of the sea and the land at La Jolla Shores, in California, by his daughter Anitra and me. You can visit with him any time in the littoral drift. Working on this edition of Waves and Beaches *refreshed memories of my own travels and research over the past forty years. I have traversed the oceans and stood on the edges of all the continents of the world; grasped sand from the beaches of the Pacific coast, Chile to Alaska. It all came alive again: bioluminescence left glittering atop berms, water seeping between grains of sand, the violence of hurricanes at sea, the chilling winds of the polar regions, and the baking Sun on Amazon mudflats off the coast of Guyana. The pulse of the sea makes a ship shudder and a surfer glide with joy, and then the wave dissolves into turbulence until the last grain of sand is at rest. Life comes in waves.*
—Kim McCoy, December 2020, La Jolla, California

Appendix A

TABLE A.1

The Progress of Wave Research as Highlighted by Milestones in the Literature

1802	Franz von Gerstner, *Theory of Waves*, Prague (now the Czech Republic) Czechoslovakia.
1825	Ernst Weber and Wilhelm Weber, "Experimental Studies of Waves," Austria.
1837	J. Scott Russell, *Report on Waves*, British Association for the Advancement of Science.
1845	George Airy, "On Tides and Waves," Encyclopedia Metropolitan, England.
1863	W. J. M. Rankine, "On the Exact Form of Waves Near the Surface of Deep Water," Royal Society of London.
1864	Thomas Stevenson, *The Design and Construction of Harbours*, England.
1877	Lord Rayleigh, "On Progressive Waves," London Mathematical Society.
1880	George G. Stokes, "On the Theory of Oscillatory Waves," England.
1904	D. D. Gaillard, *Wave Action in Relation to Engineering Structures*, US Army Corps of Engineers.
1911	Vaughan Cornish, *Waves of the Sea and Other Water Waves*, England.
1925	Harold Jeffreys, "On the Formation of Water Waves by the Wind," England.
1931	H. Thorade, *Probleme der Wasserwellen*, Germany.
1932	Horace Lamb, *Hydrodynamics*, England.
1942	Morrough P. O'Brien, "A Summary of the Theory of Oscillatory Waves," Beach Erosion Board, USA.
1947	H. U. Sverdrup and W. H. Munk, *Wind, Sea, and Swell*, Hydrographic Office, US Navy.
1955	W. J. Pierson, G. Neumann, and R. W. James, *Practical Methods for Observing and Forecasting Ocean Waves by Means of Wave Spectra and Statistics*, USA.
1963	"Ocean Wave Spectra," USN Oceanographic Office and US National Research Council Symposium.
1965	Blair Kinsman, *Wind Waves: Their Generation and Propagation on the Ocean Surface*, Englewood Cliffs, NJ: Prentice Hall.
1968	O. M. Phillips, *The Dynamics of the Upper Ocean*, Cambridge, England.
1998	Paul D. Komar, *Beach Processes and Sedimentation*, Englewood Cliffs, NJ: Prentice-Hall.
2000	David Cartwright, *Tides: A Scientific History*, Cambridge University Press.
2007	Leo Holthuijsen, *Waves in Oceanic and Coastal Waters*, Cambridge University Press.

PREVIOUS SPREAD: Go, and with your knowledge, touch each sublime wave. Bodysurfer Keith Malloy at Teahupo'o, Tahiti. *Chris Burkard*

TABLE A.2
Wind Scales and Sea Descriptions and World Meteorological Organization (WMO)
Scales in use in international reports

Beaufort Scale	Seaman's description of wind	Wind speed [knots]	Estimating wind velocities on sea	Intl. scale sea descrip. and wave heights [feet]	WMO sea state code height [meters]
0	Calm	< 1	sea like mirror	calm glassy 0	0 (zero m)
1	Light air	1 to 3	ripples no foam	rippled 0 to 1	1 (0 to 0.1)
2	Light breeze	4 to 6	small wavelets, crests have glassy appearance and do not break	smooth wavelets 1 to 2	2 (0.1 to 0.5)
3	Gentle breeze	7 to 10	large wavelets, crests begin to break, scattered whitecaps	slight 2 to 4	3 (0.5 to 1.25)
4	Moderate breeze	11 to 16	small waves become longer, frequent whitecaps	moderate 4 to 8	4 (1.25 to 2.5)
5	Fresh breeze	17 to 21	moderate waves assuming a more pronounced long form		
6	Strong breeze	22 to 27	large waves begin to form, extensive whitecaps everywhere	rough 8 to 13	5 (2.5 to 4)
7	Near gale	28 to 33	sea heaps up and white foam from breaking waves begins to be blown in streaks along the direction of the wind	very rough 13 to 20	6 (4 to 6)
8	Gale	34 to 40	moderately high waves of greater length; edges of crests break into spindrift; foam is blown in well-marked streaks along direction of the wind		
9	Strong gale	41 to 47	high waves, dense streaks of foam, spray may affect visibility, sea begins to roll		
10	Storm	48 to 55	very high waves, the surface of the sea takes on a white appearance, the rolling sea becomes heavy and shock-like, visibility affected	high 20 to 30	7 (6 to 9)
11	Violent storm	56 to 63	exceptionally high waves, small to medium-sized ships are lost from view for long periods	very high 30 to 45	8 (9 to 14)
12	Hurricane	64	the air is filled with foam and spray; sea completely white with driving spray; visibility very seriously affected	phenomenal over 45	9 (over 14)

Appendix B

Climate and Coastal Management Plans

Seychelles Coastal Management Plan: 2019–2024. World Bank; Victoria, Seychelles: Ministry of Environment, Energy and Climate Change of Seychelles.

Delta Programme 2019 Continuing the work on the delta: adapting the Netherlands to climate change in time. Ministry of Infrastructure and Water Management, Government of the Netherlands.

Isle of Wight Shoreline Management Plan 2, Appendix C: Baseline Process Understanding, C1: Annex B Climate Change and Sea Level Rise, December 2010, Coastal Management; Directorate of Economy & Environment, Isle of Wight Council.

Adapting to Rising Tides: Contra Costa County Assessment and Adaptation Project, March 2017.

The Isle of Man Climate Change Scoping Study, Technical Paper 1, Technical summary, 2006.

Building Coastal Resilience in the Face of Climate Change, Coastal Resilience Network Adaptation Guidebook, 2015, The Nature Conservancy.

Louisiana's Comprehensive Master Plan for a Sustainable Coast, June 2017, State of Louisiana.

Atlantic County Master Plan, Atlantic County, New Jersey, August 2017, Heyer, Gruel & Associates.

Coastal Erosion Planning & Response Act A REPORT TO THE 86TH LEGISLATURE, Texas General Land Office, 2018.

A Summary of Key Findings from California's Fourth Climate Change Assessment, 2018, State of California.

California Coastal Commission Sea Level Rise Policy Guidance Draft Science Update, July 2018, Chapter 3, 43–56.

The San Diego Regional Climate Collaborative, Comparing Sea Level Rise Adaptation Strategies in San Diego: An Application of the NOAA Economic Framework, June 2017.

BP Group, BP Statistical Review of World Energy, 2019, 68th edition.

Coastal Plans: "Use of Natural and Nature-Based Features (NNBF) for Coastal Resilience Final Report," Ridges, T. et al., US Army Corps of Engineers, January 2015.

"Eye on the Market, Energy Outlook 2018, Pascal's Wager," J.P. Morgan Private Bank.

Harvey, Nick, and Brian Caton. *Coastal Management in Australia*, University of Adelaide Press, South Australia, 2010. 342 pages, ISBN 978-0-9807230-3-8.

PREVIOUS SPREAD: Sea kayaking requires knowledge of ocean waves and the conditions that create them. Chris Bensch explores the tidal zone in coastal Oregon. *Fredrik Marmsater*

Further Reading

Dive deeper into oceanography including the mathematical treatment of waves:

Talley, L.D., Pickard, G.L., Emery, W.J., Swift, J.H. *Descriptive Physical Oceanography: An Introduction.* Elsevier Science, 2020, ISBN-13: 9780081009246.

Leo Holthuijsen. *Waves in Oceanic and Coastal Waters.* Cambridge University Press, 2010.

Paul D. Komar. *Beach Processes and Sedimentation.* Englewood Cliffs, NJ: Prentice-Hall, 1998.

Blair Kinsman. *Wind Waves: Their Generation and Propagation on the Ocean Surface.* 1st edition, Dover Publications, 2012.

O. M. Phillips. *The Dynamics of the Upper Ocean.* 2nd edition, Cambridge, England: Cambridge University Press, 1980.

Geology of beaches and shorelines:

Richard Davis Jr., Duncan Fitzgerald. *Beaches and Coasts.* 2nd edition, Wiley, December 2019, ISBN: 978-1-119-33455-2.

F. P. Shepard. *Submarine Geology.* New York: Harper and Row, 1973.

Shoreline engineering and naval architecture:

Robert Dean, Robert Dalrymple. *Water Wave Mechanics for Engineers and Scientists,* World Scientific, 1991, ISBN: 9789810204211.

U. S. Army Corps of Engineers. Coastal Engineering Research Center. *Shore Protection Manual.* 1984.

Harold Saunders. *Hydrodynamics in Ship Design.* New York: Society of Naval Architects & Marine Engineers, 1965.

Books featuring waves from a different point of view:

Willard Bascom. *The Crest of the Wave: Adventures in Oceanography,* 1st Anchor Books edition, (February, 1990), ISBN-13: 978-0385266338.

Peter Bruce. *Adlard Coles' Heavy Weather Sailing.* International Marine/Ragged Mountain Press; 6th edition, 2008, ISBN-13: 978-0071592901.

Joseph MacInnis. *Fitzgerald's Storm.* Macmillan, 1997.

W. G. Van Dorn. *Oceanography and Seamanship.* New York: Dodd, Mead & Co., 1974.

Waves in literature:

William Finnegan. *Barbarian Days: A Surfing Life.* Penguin, 2015.

Susan Casey. *The Wave: In Pursuit of the Rogues, Freaks, and Giants of the Ocean.* Anchor, 2011, ISBN-13: 978-0767928854.

Arthur C. Clarke. *Rendezvous with Rama.* New York: Ballantine Books, 1974.

Joshua Slocum. *Sailing Alone Around the World.* New York: Sheridan House, 1954.

Charles Darwin. *The Voyage of the Beagle.* New York: Bantam Books, 1972.

Joseph Conrad. *Typhoon.* New York: Bantam Books, 1971.

Homer. *The Odyssey.* E. V. Rieu, translation. New York: Penguin Books, 1946.

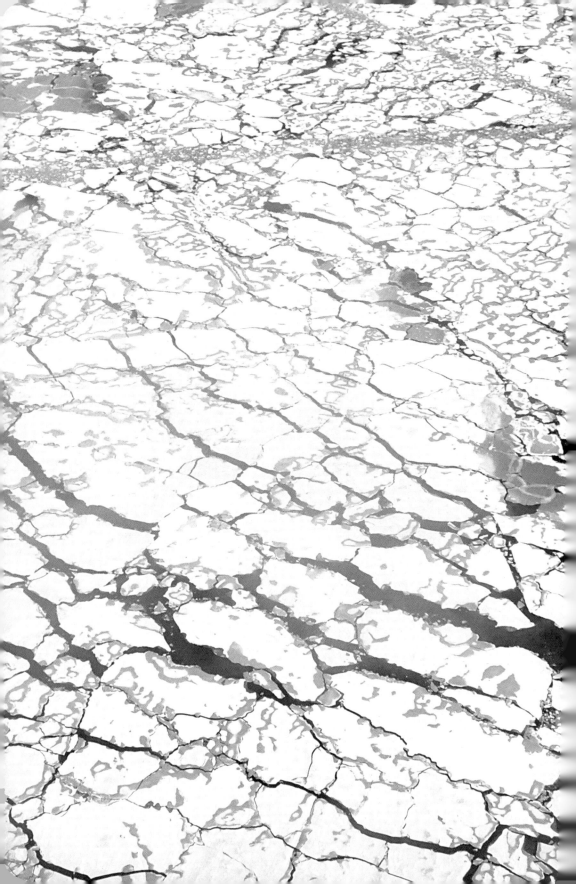

Endnotes

Chapter 1:
Genesis of Land, Water, and Waves

p.21 *The original materials...*: Charles Langmuir, *How to Build a Habitable Planet: The Story of Earth from the Big Bang to Humankind*, revised (New Jersey: Princeton University Press, 2012), 113–140, 204, 324.

p.26 *Waves range in size...*: Leo Holthuijsen, *Waves in Oceanic and Coastal Waters*, (New York: Cambridge University Press, 2010), 24–52.

p.26 *such as El Niño...*: Richard J Seymour et al., "Influence of El Ninos on California's Wave Climate," Proceedings of the 19th International Conference on Coastal Engineering, American Society of Civil Engineers, 1984, 577–592.

p.28 *arise when the drag of moving air...*: Jin Wu, "Wind-Stress Coefficients over Sea Surface near Neutral Conditions—A Revisit," *Journal of Physical Oceanography* (1980).

p.34 *ranging in size from boulders to...*: *Coastal Engineering Manual—Part III* (Washington, D.C.: Department of the Army U.S. Army Corps of Engineers, 2008), III-1-27.

p.44 *global sea level was also fairly stable...*: Kurt Lambeck et al., "Sea level and global ice volumes from the Last Glacial Maximum to the Holocene," Proceedings of the National Academy of Sciences of the United States of America, 111 (2014): 15296–15303.

p.45 *During the Younger Dryas cold snap...*: T. C. Moore Jr., "The Younger Dryas: From whence the fresh water?" *Paleoceanography*, 20 (2005), PA4021.

p.46 *An example of this post-glacial rebound...*: K. O. Emery and D. G. Aubrey, "Glacial rebound and relative sea levels in Europe from tide-gauge records," *Tectonophysics*, 120 (1985) 239–255.

p.47 *experiencing a wave of climate change...*: Special Report on the Ocean and Cryosphere in a Changing Climate (Intergovernmental Panel on Climate Change, 2019).

p.47 *within the past two centuries...*: Charles Langmuir, *How to Build a Habitable Planet: The Story of Earth from the Big Bang to Humankind*, revised (New Jersey: Princeton University Press, 2012), 555–565.

p.47 *now flow with a new wave of energy...*: Ian R. Young, Agustinus Ribal, "Multi-platform evaluation of global trends in wind speed and wave height." *Science* 364 (2019), 548–552.

Chapter 2: Ideal Waves

p.50 *The shape and motion of the ocean's surface...*: Leo Holthuijsen, Waves in Oceanic and Coastal Waters, (New York: Cambridge University Press, 2010).

p.50 *Ancient mariners knew...*: Sam Low, *Hawaiki Rising: Hokule'a, Nainoa Thompson, and the Hawaiian Renaissance* (Waipahu: Island Heritage Publishing, 2013).

p.53 *In 1802, Franz Gerstner...*: U.S. Army Corps of Engineers, *Shore Protection Manual* (Washington, D.C., 1984), volume I, chapters 1–5.

p.56 *depends upon wave period...*: *Coastal Engineering Manual—Part II* (Washington, D.C.: U.S. Army Corps of Engineers, 2015), volume II, 1–22.

PREVIOUS SPREAD: The Canadian icebreaker *Louis S. St-Laurent* and the USCGC *Healy* tie up during a joint oceanographic research project in the warming Arctic Ocean. *Patrick Kelley/US Coast Guard*

Chapter 3: Wind Waves

p.79 *The North Atlantic near Iceland...*: H. Santo, P. H. Taylor, R. Gibson, "Decadal variability of extreme wave height representing storm severity in the northeast Atlantic and North Sea since the foundation of the Royal Society," *Philosophical Transactions of the Royal Society A, Mathematical, Physical, and Engineering Sciences,* 472 (2016), 20160376.

p.81 *sometimes called extreme waves...*: Paul Stansell, "Distributions of extreme wave, crest and trough heights measured in the North Sea," *Ocean Engineering,* 32, 17-18, (2005), 2241-2242.

p.85 *are over 2 terawatts continuously...*: "An Introduction to Internal Waves" Theo Gerkema, J. T. F. Zimmerman, Netherlands Institute for Sea Research (2008), 9-22.

p.88 *Professor and mariner...*: Charles S. Cox, Xin Zhang, and Timothy F. Duda, "Suppressing breakers with polar oil films: Using an epic sea rescue to model wave energy budgets," *Geophysical Research Letters,* 44 (2017), 1414-1421.

p.89 *kelp (large brown algae)...*: K. Filbee-Dexter, C. J. Feehan, R. E. Scheibling, "Large-scale degradation of a kelp ecosystem in an ocean warming hotspot," *Mar Ecol Prog Ser,* 543 (2016), 141-152.

p.89 *sea ice also reduce wave action...*: Alex DeMarban, "Another Bering Sea storm threatens Western Alaska with flooding and major erosion," *Anchorage Daily News* (2017).

p.90 *what is called "pancake ice"...*: Jim Thompson et al., "The Balance of Ice, Waves, and Winds in the Arctic Autumn," *Eos,* 98 (2017).

p.90 *detailed charts of the Arctic...*: Perry, R.K., et al., *Bathymetry of the Arctic Ocean* (Boulder, Colorado: Geological Society of America, 1986).

p.93 *arriving at the US Pacific coast...*: Peter Ruggiero, Paul D. Komar, Jonathan C. Allan, "Increasing wave heights and extreme value projections: The wave climate of the U.S. Pacific Northwest," *Coastal Engineering,* 57 (2010), 539-552.

p.94 *forecast the arrival of waves...*: *Guide to Wave Analysis and Forecasting,* second edition (Geneva, Switzerland: Secretariat of the World Meteorological Organization, 1998).

Chapter 4: Waves in Shallow Water

p.98 *the entire range of wave...*: *Coastal Engineering Manuals* (Washington, D.C.: U.S. Army Corps of Engineers, 2002-2015), parts 1-4.

p.98 *react in special ways...*: Edward and B. Thornton. R. T. Guza, "Transformation of Wave Height Distribution," *Journal of Geophysical Research,* 88 (1983), 5925-5938.

p.105 *sea caves resonate with the wave's energy...*: Two examples of wave-activated resonant acoustic sculptures are: The Wave Organ in San Francisco (by Peter Richards and George Gonzales, and inspired by Bill Fontana) and the Sea Organ in Zadar, Croatia (by Nikola Bašić)

p.110 *site of Lothal...*: Brian Fagan, *The Attacking Ocean: The Past, Present, and Future of Rising Sea Levels,* (Bloomsbury Press, 2014), 117.

p.112 *tropical cyclone Bhola...*: N. Hossain, "The 1970 Bhola cyclone, nationalist politics, and the subsistence crisis contract in Bangladesh," *Disasters,* 4 (2017).

p.113 *Katrina affected the entire...*: H. Fritz et al., "Hurricane Katrina storm surge distribution and field observations on the Mississippi Barrier Islands," *Estuarine, Coastal and Shelf Science,* 74 (2007), 12-20.

p.114 *Hurricane Sandy struck...*: Ning Lin et al., "Hurricane Sandy's flood frequency increasing from year 1800 to 2100," *Proceedings of the National Academy of Sciences of the United States of America* 113:43 (2016).

p.115 *Super Typhoon Haiyan...*: M. Bayani Cardenas et al., "Devastation of aquifers from tsunami-like storm surge by Supertyphoon Haiyan," *Geophysical Research Letters,* 42 (2015).

p.116 *risk deteriorating into chaos...*: Mitchell Waldrop, *Complexity: The Emerging Science at the Edge of Order and Chaos* (New York: Simon & Schuster, 1992).

Chapter 5:
Winds and Waves of Climate Change

p.120 *others transform endlessly, forever...*: Renêe Hetherington, Robert G. B. Reid, *The Climate Connection: Climate Change and Modern Human Evolution* (British Columbia: Cambridge University Press, 2010).

p.121 *complex and far reaching...*: Malcolm H. Wiener, "The Interaction of Climate Change and Agency in the Collapse of Civilizations ca. 2300–2000 BC," Tree-Ring Research 70(3), (2014), S1–S16.

p.121 *healthy coral reef depends...*: Simon D. Donner et al., "Global assessment of coral bleaching and required rates of adaptation under climate change," *Global Change Biology*, 11 (2005), 2251–2265.

p.121 *spur and groove alignment...*: Walter H. Munk, Marston C. Sargent, "Adjustment of Bikini atoll to ocean waves," *Trans Am Geophys Union*, 29 (1948), 855–860.

p.122 *surrender through erosion...*: Ellen Quataert et al., "The influence of coral reefs and climate change on wave-driven flooding of tropical coastlines," *Geophysical Research Letters*, 42 (2015), 6407–6415.

p.122 *for more than a decade...*: Darren James, "Lost at sea: The race against time to save the Carteret Islands from climate change," *The World*, Australian Broadcasting Corporation (ABC), August 3, 2018.

p.122 *Including Boston...*:" City of Boston website, "Resilient Boston Harbor," (2018), https://www.boston.gov/departments/environment/resilient-boston-harbor.

p.122 *must confront the rising seas...*: Gary Griggs et al., "Rising Seas in California: An Update on Sea-Level Rise Science" (California Ocean Science Trust, 2017).

p.122 *provide a national policy...*: Howard Kunreuther, "Reauthorizing the National Flood Insurance Program," *Issues in Science and Technology*, National Academies of Science 34:3 (2018).

p.122 *sea level changes...*: Britt Raubenheimer et al. "The Nearshore Water-Land System During Major Storms," Conference: International Conference on Coastal Sediments (2019), 13–24.

p.123 *being maturely addressed...*: Robert E. Kopp et al., "Usable science for managing the risks of sea-level rise," *Earth's Future*, 7 (2019), 1235–1269.

p.123 *the Isle of Wight...*: *Isle of Wight Shoreline Management Plan 2* (Coastal Management, Isle of Wight Council, Climate Change and Sea Level Rise, 2010).

p.123 *sea ice is decreasing...*: Robert A. Massom et al., "Antarctic ice shelf disintegration triggered by sea ice loss and ocean swell," *Nature* 558 (2018), 383–389, and Yan Liu et al., "Iceberg calving of Antarctic ice shelves," *Proceedings of the National Academy of Sciences*, 112:11 (2015), 3263–3268.

p.125 *it will be underwater...*: *California Coastal Commission Sea Level Rise Policy Guidance Final Adopted Science Update* (California Coastal Commission, 2018), Chapter 3: 43–56.

p.127 *supply us with sea level rise...*: S. Jevrejeva et al., "How will sea level respond to changes in natural and anthropogenic forcings by 2100?" *Geophysical Research Letters*, 37 (2010).

p.127 *the Law of the Sea...*: E. Borghese, "The Law of the Sea," *Scientific American* 248:3 (1983), 42–49.

p.127 *new international patent law...*: "Technologies or Applications for Mitigation or Adaptation Against Climate Change" (European Patent Office, Global Cooperative Patent Classification Y02, 2018).

p. 127 *wave of human population...*: Vaclav Smil, *Growth: From Microorganism to Megacities* (MIT Press, 2019).

Chapter 6: Tides and Seiches

p.132 *beyond the scope of this book...*: David Edgar Cartwright, *Tides: A Scientific History* (Cambridge University Press, 2002).

p.133 *longest waves oceanographers...*: K. Wunsch et al., "Dynamics of long-period tides," *Progress in Oceanography*, 40:1 (1997), 81–108.

p.133 *Not until a number of explorers...*: D. Cartwright, "On the Origins of Knowledge of the Sea Tides from Antiquity to the Thirteenth Century," *Earth Sciences History: 2001*, 20:2 (2001), 105–126.

p.133 *some 1,500 years passed before...*: B. van der Waerden, "The Heliocentric System in Greek, Persian and Hindu Astronomy," *Annals of the New York Academy of Sciences*, 500:1 (1987), 525–545.

p.143 *visible wave front or bore exists...*: D. Lynch, "Tidal Bores," *Scientific American*, 247 (1982), 146–156.

p.146 *winds across the Baltic Sea...*: E. Kulikov et al., "Variability of the Baltic Sea Level and Floods in the Gulf of Finland," *Oceanology*, 53:2 (2013), 145–151.

p.146 *inundations of St. Petersburg...*: B. Jönsson. et al, "Standing waves in the Gulf of Finland and their relationship to the basin-wide Baltic seiches" *Journal of Geophysical Research Atmospheres*, 113 (2008).

p.152 *while studying surging in...*: J. Park et al., "Water level oscillations in Monterey Bay and Harbor," *Ocean Science Discussions*, 11:6 (2014), 2569–2606.

Chapter 7: Impulsively Generated Waves

p.159 *west coast of Sumatra...*: K. Monecke, "A 1,000-year sediment record of tsunami recurrence in northern Sumatra," *Nature*, 455 (2008), 1232–1234.

p.159 *measured 9.1 Richter...*: US Geological Survey website, "Northern Sumatra M9.0 - December 26, 2004," https://earthquake.usgs.gov/learn/topics/sumatra9.php.

p.166 *global array of seismic...*: L. Zerbo, "Sensing the Danger: Can Tsunami Early Warning Systems Benefit from Test Ban Monitoring?" (International Atomic Energy Agency, 2005), 47-1.

p.170 *destruction due to the tsunami...*: S. Violette, G. Boulicot, S. M. Gorelick, "Tsunami-induced Groundwater Salinization in Southeastern India"

Comptes Rendus Geoscience, 341:4 (2008), 339–346.

p.170 *the nuclear reactors...*: M. Sugiyama et al., "Research management: Five years on from Fukushima," *Nature*, 531 (2016), 29–31.

p.171 *development in renewable energy...*: Fukushima Minpo, "Fukushima unveils grand plan for alternative energy transmission line networks" *Japan Times*, 2016.

p.171 *caused financial, societal, and...*: Editor: Richard Hindmarsh, *Nuclear Disaster at Fukushima Daiichi: Social, Political and Environmental Issues*, Routledge Studies in Science, Technology and Society, NY, April 2013, pp. 1–18, ISBN: 978-0-203-55180.

p.171 *the original report...*: J. H. Latter, "Tsunamis of volcanic origin: Summary of causes, with particular reference to Krakatoa, 1883," *Bulletin Volcanologique*, 44, 3, (1981), 467–490.

p.173 *small island began to emerge...*: T. Giachetti et al., "Tsunami hazard related to a flank collapse of Anak Krakatau Volcano, Sunda Strait, Indonesia," Geological Society London Special Publications 361 (2012), 79–90.

p.174 *many civilizations weakened...*: M. Wiener, "The Interaction of Climate Change and Agency in the Collapse of Civilizations ca. 2300-2000 BC," *Radiocarbon*, 70 (2014), 1–16.

p.174 *the eastern Mediterranean...*: D. Langgut et al., "Climate and the Late Bronze Collapse: New Evidence from the Southern Levant," *Journal of the Institute of Archaeology of Tel Aviv*, 40:2 (2013), 149–175.

p.174 *was never the same after...*: Robert Drews, *The End of the Bronze Age* (New Jersey: Princeton University Press, 1993).

p.174 *collapsed into the sea...*: Hermann M. Fritz, "Initial phase of landslide generated impulse waves," Swiss Federal Institute of Technology Zürich (2002), B-22-25.

p.176 *was a different situation...*: W. G. Van Dorn, Ivy Mike: *The First Hydrogen Bomb* (Xlibris Corporation, 2008).

p.177 *radiating from a ship...*: M. McKenna et al., "Underwater radiated noise from modern commercial ships," *Journal of the Acoustical Society of America*, 131 (2012), 92.

p.179 *theoretical calculations and computational...*: E. Campana et al., "Shape optimization in ship hydrodynamics using computational fluid dynamics," *Methods Appl. Mechanical Engineering*, 196 (2006) 634–651, and J. R. Thomas et al., *Computer Methods in Applied Mechanics and Engineering*, 196:1–3 (2006), 634–651.

Chapter 8: Measuring and Making Waves

p.189 *nearshore sediments are measured...*: P. Ruggiero et al., "National assessment of shoreline change—Historical shoreline change along the Pacific Northwest coast," *U.S. Geological Survey Open-File Report* (2012), 1007, 62.

p.196 *in a floating buoy...*: W. Schmidt et al., "A GPS-tracked surfzone drifter," *J. Atmos. Oceanic Technol.* 20 (2003), 1069–1075.

p.198 *Airborne Laser Bathymetry...*: G. Guenther et al., Airborne laser hydrography: System design and performance factors, (Rockville, Maryland: U.S. Dept. of Commerce, National Oceanic and Atmospheric Administration, National Ocean Service, Charting and Geodetic Services, 1985), 385, and G. Guenther et al., Meeting the Accuracy Challenge in Airborne Bathymetry, (Dresden, Germany: European Association of Remote Sensing Laboratories 2000), 23.

p.198 *surveying of the coastal...*: J. Irish et al., "Coastal engineering applications of high-resolution lidar bathymetry" *Coastal Engineering*, 35:1-2 (1998), 47–71.

p.198 *Lidar can accurately...*: K. Martins et al., "High-resolution monitoring of wave transformation in the surf zone using a LiDAR scanner array," *Coastal Engineering*, 128 (2017), 37–43.

p.198 *image of the coastline...*: K. Martins et al., "High-resolution monitoring of wave transformation in the surf zone using a LiDAR scanner array," *Coastal Engineering*, 128 (2017), 37–43.

p.198 *optical backscatter sensor (OBS), revolutionized...*: J. P. Downing, R. W. Sternberg, and C. Lister, "New Instrumentation for the Investigation of Sediment Suspension Processes in the Shallow Marine Environment," *Marine Geology*, 42 (1981), 19–34.

p.200 *satellites equipped with altimeters...*: M. Ablain et al., "Improved sea level record over the satellite altimetry era (1993–2010)," *Ocean Science*, 11 (2015), 67–82.

p.202 *up into the swash...*: R. Guza and E. B. Thornton, "Swash oscillations on a natural beach," *Journal of Geophysical Research*, 87 (1982), 483–491.

p.203 *impact of a breaking wave...*: G. Cuomo et al., "Breaking wave loads at vertical seawalls and breakwaters," *Coastal Engineering*, 57:4 (2010), 424-439.

p.204 *observable in many small-scale oceanic...*: O'Donnell, Michael, and Denny, Mark. (2008). "Hydrodynamic Forces and Surface Topography: Centimeter-Scale Spatial Variation in Wave Forces." *Limnology and Oceanography*, 53, 579–588. 10.2307/40006442.

p.204 *withstand breaking waves...*: Thornton, Wu, and Guza, "Breaking Wave Design Criteria," Coastal Engineering Proceedings, 19 (1984).

p.208 *atmospheric gases...*: Christoph Garbe et al., "Transfer across the Air-Sea Interface: Ocean-Atmosphere Interactions of Gases and Particles," *Springer* (2013) 55–112.

Chapter 9: The Surf

p.225 *World War II Waves Project...*: Steve Barilotti, "Beach Party—Willard Bascom, The War Department, and the Wildcat Origins of Modern Surf Science," *The Surfer's Journal*, 24(6) 90–99.

p.233 *the first Westerner to...*: K. Eschner, "What the First European to Visit Hawaii Thought About Surfers," *Smithsonian*, 2017.

p.235 *best waves for surfing...*: N. Pizzo, "Surfing surface gravity waves," *Journal of Fluid Mechanics*, 823 (2017), 316–328.

p.237 *have all of these features...*: E. Bradley et al., "Sustainable Management of Surfing Breaks: Case Studies and Recommendations," *J. of Coastal Research*, 253 (2009), 684–703, and *Coastal Engineering Manual—Part II* (Washington, D.C.: U.S. Army Corps of Engineers, 2015), II-2-56-67.

Chapter 10: Beaches: Where the Surf Meets the Sediment

p.249 *beaches vary widely...*: Arjen Luijendijk et al., "The State of the World's Beaches," *Nature, Scientific Reports*, 8 (2018), 6641.

p.250 *beach responds with great sensitivity...*: Nicole Elko et al., "The future of nearshore processes research," *Shore & Beach*, 83:1 (2015), 13–38.

p.259 *Fort Ord (near Carmel) has been studied many...*: T. Keefer, "Dune erosion, mega-cusps and rip currents modeling of field data," (M.Sc. Thesis, Naval Postgraduate School).

p.260 *the wave's energy shakes...*: Adam P. Young et al., "Coastal cliff ground motions from local ocean swell and infragravity waves in southern California," *Journal of Geophysical Research*, 116 (2011), C09007.

p.260 *back into the depths...*: O. B. Wilson, Jr., "Measurements of acoustic ambient noise in shallow water due to breaking surf" (Naval Postgraduate School, 1982).

p.263 *effects of wave energy...*: Brian C. McFall, "The Relationship between Beach Grain Size and Intertidal Beach Face Slope," *Journal of Coastal Research*, 35(5) p. 1080–1086.

p.269 *cusps are evenly spaced...*: *Coastal Engineering Manual—Part III* (Washington, D.C.:, Department of the Army U.S. Army Corps of Engineers, 2008), III-2-49-57.

p.275 *beaches of the world are strewn...*: L. Eriksen et al., "Plastic Pollution in the World's Oceans: More than 5 Trillion Plastic Pieces Weighing over 250,000 Tons Afloat at Sea," *PLOS ONE* 9:12 (2014), e111913.

p.275 *not all plastic debris floats...*: D. Cressey, "Bottles, bags, ropes and toothbrushes: the struggle to track ocean plastics," *Nature*, 536 (2016), 265.

p.277 *resolution was signed...*: John Ndiso, "Nearly 200 nations promise to stop ocean plastic waste," Reuters, December 6, 2017.

Chapter 11: The Conveyor Belts of Sand

p.288 *coast from Point Conception...*: *Coastal Regional Sediment Management Plan Central Coast from Pt. Conception to Pt. Mugu: Final Report* (Beach Erosion Authority for Clean Oceans and Nourishment, 2009).

p.288 *was funded by NSTS...*: R. J. Seymour (editor), *Nearshore Sediment Transport* (Boston: Springer, 1989).

p.299 *an ancient tsunami inundation...*: C. Peterson et al., "Paleotsunami Inundation of a Beach Ridge Plain: Cobble Ridge Overtopping and Interridge Valley Flooding in Seaside, Oregon, USA," *Journal of Geological Research*, (2010), 276989.

p.300 *In the 1970s...*: S. L. Costa and J. D. Isaacs, "The Modification of Sand Transport in Tidal Inlet Inlets" (Charleston, S.C.: Coastal Sediments 77, Fifth Symposium of the Waterway, Port Coastal and Ocean Division, American Society of Civil Engineers, 1977).

p.300 *interacts with the bed forms...*: *Coastal Engineering Manual—Part II* (Washington, D.C.: Department of the Army U.S. Army Corps of Engineers, 2008), III-2-1-15.

p.305 *were nourished in...*: R. Seymour et al., "Rapid erosion of a small Southern California beach fill," *Coastal Engineering*, 52 (2005), 151–158.

p.305 *and Oceanside for a total...*: M. Grubbs, "Beach Morphodynamic Change Detection using LiDAR during El Niño Periods in Southern California" (M.Sc. Thesis, University of Southern California, 2017).

p.306 *growth and climate change...*: P. Adams, D. Inman, and J. Lovering, "Effects of climate change and wave direction on longshore sediment transport patterns in Southern California," *Springer Climatic Change*, 109:1 (2011), 211–228.

p.306 *All these have changed...*: Cope M. Willis and Gary B. Griggs, "Reductions in Fluvial Sediment Discharge by Coastal Dams in California and Implications for Beach Sustainability," *The Journal of Geology*, 111 (2003), 167–182.

p.307 *methods that already exist...*: R. Dean and R. Dalrymple, *Coastal Processes with Engineering Applications* (Cambridge University Press, 2002), part four, 341–464.

p.307 *implement new local and national policies...*: Howard Kunreuther, "Reauthorizing the National Flood Insurance Program," *Issues in Science and Technology*, National Academies of Science 34:3 (2018).

p.308 *diminish the supply of sediments...*: E. Ezcurra et al., "A natural experiment reveals the impact of hydroelectric dams on the estuaries of tropical rivers," *Science Advances*, 5:3 (2019), eaau9875.

p.308 *problems associated with the building of dams...*: *Dams and Development*: The Report of The World Commission on Dams (Earthscan Publications Ltd, 2000).

p.308 *shoreline erosion continues...*: M. Hereher, "Vulnerability of the Nile Delta to sea level rise: an assessment using remote sensing Geomatics," *Natural Hazards and Risk*, 1:4 (2010).

p.309 *Three Gorges Dam...*: Z. Yang, "Dam impacts on the Changjiang (Yangtze) River sediment discharge to the sea: The past 55 years and after the Three Gorges Dam," *Water Resources Research*, 42 (2006), W04407.

p.310 *funded the Amazon Shelf...*: C. Nittrouer et al., "AMASSEDS: An Interdisciplinary Investigation of a Complex Coastal Environment," *Oceanography*, 4:1 (1991), 3–7.

p.310 *tons of river sediments...*: G. Kineke et al., "Fluid-mud Processes on the Amazon Continental Shelf," *Continental Shelf Research*, 16:5-6 (1996), 667–696.

p.310 *what happens thousands of miles from the ocean...*: "Sand, rarer than one thinks" (United Nations Environment Programme, 2014).

p.310 *visible at many river deltas...*: L. Van Rijn, "Manual Sediment Transport Measurements in Rivers, Estuaries and Coastal Seas," *Aquapublications*, (2007), 500.

p.313 *across the Gaoping River...*: J. Liu et al., "The effect of a submarine canyon on the river sediment dispersal and inner shelf sediment movements in southern Taiwan," *Marine Geology*, 181:4 (2002), 357–386.

p.313 *balance of sediments and water...*: M. Kondolf, "Effects of Dams and Gravel Mining on River Channels," *Environmental Management*, 21:4 (1997), 533–551.

p.313 *collapsed, killing...*: J. J. Sousa and L. Bastos, "Multi-temporal SAR interferometry reveals acceleration of bridge sinking before collapse," *Nat. Hazards Earth Syst. Sci.* 13 (2013), 659–667.

p.313 *hope in some areas...*: K. Horner, "Dams be damned, let the world's rivers flow again," *The Guardian*, January 9, 2017.

p.313 *Elwha and the Glines...*: A. Ritchie, "Morphodynamic evolution following sediment release from the world's largest dam removal," et al., *Scientific Reports*, 8, 13279 (2018) doi:10.1038/s41598-018-30817-8

p.314 *equilibrium, is easily upset...*: T. Wamsley et al., "Guidance for Developing Coastal Vulnerability Metrics," *Journal of Coastal Research*, 31:6 (2015), 1521–1530.

p.314 *regions of the world's deltas...*: J. Reker et al., "Deltas on the move: Making deltas cope with the effects of climate change," Dutch National Research Programmes Climate Changes Spatial Planning, 2006.

p.314 *decades of satellite data...*: J. Coleman and Oscar K. Huh, "Major deltas of the world: a perspective from Space" (Baton Rouge, Louisiana: Coastal Studies Institute, Louisiana State University, 2004), 74.

p.314 *accelerated sinking rates...*: James P.M. Syvitski et al., "Sinking Deltas," *Nature Geoscience*, 2:10 (2009).

p.314 *at 90 percent and the Nile River...*: Elham Ali and Islam A. El-Magd, "Impact of human interventions and coastal processes along the Nile Delta coast, Egypt during the past twenty-five years," *The Egyptian Journal of Aquatic Research*, 42:1 (2006) 1–10.

p.315 *areas are in retreat...*: Edward Anthony et al., "Linking rapid erosion of the Mekong River delta to human activities," *Nature Scientific Reports*, 5 (2015), 14745.

p.316 *removal of mangroves...*: M. Spalding et al., "Mangroves for coastal defence. Guidelines for coastal managers & policy makers," Wetlands International and The Nature Conservancy (2014), 42.

p.316 *gravel and sand mining...*: C. Vörösmarty et al., "Anthropogenic sediment retention: major global impact from registered river impoundments," *Global and Planetary Change*, 39 (2003), 169–190.

p.316 *Delta in Decline...*: J. Jackson and S. Chapple, *Breakpoint: Reckoning with America's Environmental Crises* (New Haven: Yale University Press, 2018).

Chapter 12: Man Against the Sea

p.334 *coastal defense is the sand dune...*: H. Dethier et al., "Design with Nature strategies for shore protection—The construction of a cobble berm and artificial dune in an Oregon State Park," U.S. Geological Survey Scientific Investigations Report 2010, Proceedings of a State of the Science Workshop, 117–126.

p.340 *washed away during...*: M. Kirkgoz, "Shock pressure of breaking waves on vertical walls," *Journal of Waterway, Port, Coastal and Ocean Division*, 108 (1982), 81–95.

p.340 *inspections, and maintenance will continue...*: J. Bottin et al., "Periodic Inspection of Humboldt Bay Jetties" (Washington, D.C.: U.S. Army Corps of Engineers, 1997).

p.342 *Rijkswaterstaat authority...*: J. Lonnquest et al., "Two Centuries of Experience in Water Resources Management: A Dutch-U.S. Retrospective" (Institute for Water Resources, U.S. Army Corps of Engineers and Rijkswaterstaat, Ministry of Infrastructure and the Environment, 2014).

p.342 *path that the world must travel...*: M. Kimmelman, "The Dutch Have Solutions to Rising Seas. The World Is Watching," *New York Times*, June 15, 2017.

p.345 *Hornsea 1 project...*: "Construction begins for UK's 1.2GW Hornsea Project One windfarm," *Power Technology*, January 29, 2018.

p.346 *the East Med pipeline...*: "H. R. 1865, December 2019, Division J Foreign Policy, TITLE II—Eastern Mediterranean Security and Energy Partnership," *Public Law*, 116-94, December 20, 2019.

p.350 *are still frequent...*: R. George, "Worse things still happen at sea: the shipping disasters we never hear about," *The Guardian*, January 10, 2015.

p.350 *Each year 1,000+ shipping...*: J. Cave, "Thousands Of Containers Fall Off Ships Every Year. What Happens To Them?" The Huffington Post, July 17, 2014.

p.351 *containers are lost...*: "Containers Lost At Sea—2017 Update" (World Shipping Council, 2017).

p.354 *inspections by insurance companies...*: "Shipping losses lowest this century, but incident number remain high," Allianz Global Corporate & Specialty Press Release, June 4, 2019.

p.355 *endurance against waves...*: Sue Neales, "'Huge' Southern Ocean swells test Volvo round-the-world racers," *The Australian*, December 27, 2017.

p.358 *explosion of the Deepwater Horizon...*: "Statement of Principles and Convening Order Regarding Investigation Into the Marine Casualty, Explosion, Fire, Pollution, and sinking of Mobile Offshore Drilling unit Deepwater Horizon, with loss of life in the Gulf of Mexico, 21–22 April 2010," Joint Department of the Interior and Department of Homeland Security, and "Deepwater Horizon Marine Casualty Investigation Report" (Republic of the Marshall Islands, Office of the Maritime Administrator, 2011).

p.358 *legally very complex...*: R. Richards, "Deepwater Mobile Oil Rigs in the Exclusive Economic Zone and the Uncertainty of Coastal State Jurisdiction,"

Journal of International Business and Law 10:2 (2011), Article 10.

p.360 *anthropogenic CO₂...*: C. Le Quere et al., "The global carbon budget 1959–2011," *Earth Syst. Sci. Data*, 5 (2013), 165–185.

p.360 *allowable sulfur...*: "CO₂ Emissions from International Maritime Shipping," United Nations UNEP DTU Partnership Working Paper 2017:4.

p.360 *hydrocarbon energy sources...*: Vaclav Smil, *Energy and Civilization: A History* (MIT Press, 2017).

p.360 *spills on the waves...*: Jonathan Ramseur, "Oil Spills: Background and Governance," Library of Congress Congressional Research Service, 2017.

p.360 *clumps of oil...*: Arne Jernelöv, "The Threats from Oil Spills: Now, Then, and in the Future," *Ambio*, 39:5–6 (2010), 353–366.

p.361 *the winds and waves...*: H. Lamb, "The Weather of 1588 and the Spanish Armada," *Royal Meteorological Society*, 43:11 (1988), 386–395.

p.361 *sunk far more vessels...*: Polybius, *The Rise of the Roman Empire* (New York: Penguin Books, 1979), 81–84.

p.361 *armed conflicts or terrorist...*: The Battles of Jutland (1916) and Lyte Gulf (1944) were the greatest naval exchanges in the 20th century. The maritime operations in Normandy (1944) were the largest in history exceeding those of the second Punic War (2,200 years ago), yet the strength of wind, height of waves, and phase of tides each played intimate roles in victory and defeat.

At this moment, somewhere, a storm wind blows, waves cross oceans, and breakers collide with distant shores. Grand Harbour, Valletta, Malta. *Kurt Arrigo*

Glossary

Selected Nautical, Coastal, and Astronomical Terms

Albedo: the fraction of light that Earth reflects back into space compared to the amount absorbed; it is fundamental to the flow of energy from the Sun into the Earth's physical and biological systems.

Apogee: the furthest point attained by the Moon, celestial object, or other satellite during its orbit.

Bar: a ridge of sand, usually underwater and parallel to the beach, sometimes exposed during low tides.

Barometric: refers to atmospheric pressure, a decrease of 1 millibar pressure causes the local sea surface to rise by 1 centimeter.

Barrier Island: a coastal-fringing landform created by the deposition of sediments over time.

Barycenter: the center of rotation around which the Earth-Moon (or other) system rotate.

Bathymetry: the measure of underwater depths, similar to the measurement of heights (topography) on land.

Beach Face: seaward side of the berm against which the waves are in constant contact.

Berm: the nearly horizontal portion of a beach, usually above-water and formed by the deposition of sediments during high tides and storm waves.

Bow: the front of a vessel during the normal direction of travel (the direction is called forward or ahead).

Breaker: a wave at its maximum steepness having become turbulent at its crest (spilling, plunging, or collapsing).

Capillary Wave: the smallest type of wave, controlled by the force of surface tension.

Clapotis: when waves are reflected back in a seaward direction from a vertical surface creating a standing wave.

Deep-Water Waves: waves traveling in water deeper than half of their wave length.

Doldrums: region near the equator with light or no winds as a result of atmospheric circulation patterns.

Doppler Shift: the apparent change in frequency of waves (sound or electromagnetic) due to the relative motions of source and receiver (toward = higher, away = lower).

Drowned Topography: landforms now submerged because of a prior rise in the local relative sea level.

DUKW (pronounced "duck"): Amphibious, six-wheeled truck with propeller and rudder used in surf zone research in big waves (D – model year 1942, U – amphibious 2½ ton truck, K – all wheel drive, W – dual rear axles). It was developed in secret during WWII, used in Sicily in 1943, Normandy in 1944, and throughout the Pacific.

PREVIOUS SPREAD: Ancient Polynesians used their intimate knowledge of waves and ocean conditions to expand across the Pacific and colonize the geographical area between Aotearoa (New Zealand), Rapa Nui, and Hawai'i. Kaua'i, Hawai'i. *Rui Camilo*

Ecliptic of the Earth: the geometric plane of the orbit of the Earth around the Sun.

Edge Wave: a special type of shallow-water wave that when reflected off a beach, is then refracted back to the beach.

Fathom: a measure of depth equal to 6 feet (2 yards or 1.82 meters).

Fathometer: an instrument used to determine water depths.

Fetch: the distance over which winds blow during the creation of waves.

GM: a ship's metacentric height measured in distance (units of length).

Gravity Waves: the most familiar type of wave, in which gravity is the restoring force.

Gyre: a large current of water that rotates as it moves, and similar in its motion to a much smaller swirling eddy or whirlpool.

Hindcast: the procedure of looking at historical data to reconstruct what contributed to a past event.

Hydrodynamics: the study of the flows and forces of water as they act upon solid bodies.

Hydrological Cycle: the path of water as it is evaporated off bodies of water to form atmospheric gas, condensed into liquid water and falling as rain which then fills rivers, replenishes aquifers, and flows into the seas; there, the heat of the Sun evaporates the water and it rises back into the atmosphere to complete the cycle. The minerals dissolved in water do not evaporate, they remain behind at the end of each cycle, and it is for this reason that the ocean has become salty.

Kinetic Energy: the energy of an object related to its mass and speed (relative to a frame of reference).

King Tide: common (non-scientific) term for an exceptionally large tide created by the alignment of the Moon, Earth, and Sun.

Lee: away from or protected from the wind, sheltered (windward is the opposite direction of leeward).

Littoral Transport: the movement of sediments along a coast driven primarily by wave energy.

Local Sea Level: the level of the sea at one location if left undisturbed by changing waves, tides, winds, and barometric pressure.

Mass Transport: the movement of water or sediments due to wave action or ocean currents.

Mean Sea Level (MSL): the mean height of the water level (over a nineteen-year period of measurements). MSL is a legal reference for marine navigation and property lines in many jurisdictions.

Mean Lower Low Water (MLLW): the nineteen-year average of the "lower low water" (lowest tide of the day) and used as the reference point (for depths and heights) for nautical charts in the United States.

Neap Tides: smaller tides created when the gravitational forces of Moon and Sun are at right angles.

Orthogonal: a line drawn at a right angle to the depth contour line, used to calculate wave energy propagation.

Oscillatory: repeatedly moving up and down, back and forth, or both (circular).

Perigee: the nearest point attained by the Moon, celestial object, or other satellite during its orbit.

Physiographic: the physical shapes and patterns of landforms and their geomorphologies.

Port: the left side of a vessel (when looking in the direction of travel); starboard is the right side.

Post-Glacial Rebound: the upward motion of the land (relative to sea level) in response to a melted glacier's decreased weight.

Potential Energy: the energy of an object related to its mass and height above a reference point.

Refraction Coefficient: in shallow water, the amount of bending and re-focusing of waves as they slow, change direction, and redirect their energy.

Rip Current: an offshore-flowing current, created by surf zone wave actions and occurring usually between two offshore bars.

Saltation: the jumping motion of a grain of sand as it is lifted by turbulence and then moved by the current.

Sea: shorter-period waves recently formed by local winds.

Scarp: a steep, nearly vertical feature, usually on the face of a berm, which has been cut back by waves.

Seiche: a form of a reflected wave that sloshes back and forth in a partially or fully enclosed basin.

Shallow-Water Waves: waves traveling in water that is shallower than half of their wave length.

Shear: the sliding and deformation that occurs in a fluid due to a difference in relative speed.

Ship Type: an indicator of a ship's function (such as MV = motor vessel, RV = research vessel, SS = steam ship).

Shoal: (process) the effect that shallow water has on wave steepness, speed, and direction (i.e., shoaling).

Shoal: (geographic) a location where the process of shoaling occurs; also describes a shallow area of water.

Spring Tides: larger tides created when the gravitational forces of Moon and Sun are aligned such as when the Moon is full or new.

Stern: the rear end of a vessel (the direction is referred to as astern).

Surf Zone: area extending from where waves break, shoreward to the top of the swash on the beach face.

Surface Tension: a property of liquids that pulls molecules together; important for capillary waves and the form of every drop of water.

Swash: the upwash and backwash movements (and location) of water on the sloping portion of a beach face.

Swell: longer-period, smoother-looking waves; usually generated by a distant storm.

TEU: Twenty-foot Equivalent Unit container size, 20 feet long, 8 feet wide, and 8.5 feet high (or 33 cubic m).

Tombolo: an island that is attached by sediments (a bar or spit) to the mainland.

Tsunami: a seismic sea wave, "harbor wave," from the Japanese tsu for "harbor" and nami for "wave."

Wave of Translation: when a wave's movement of water is mostly horizontal, as with a tsunami, tidal bore, or water over a shallow bar.

Wave Steepness: a wave's height-to-length ratio (H/L); maximum steepness is approximately 1 to 7.

Wind blowing offshore will cause the face of a wave to streak upward then tear off into a rain-like spray. These droplets (water and salt aerosols) increase the exchange of heat between the ocean and atmosphere. Pipeline, O'ahu. *Brian Bielmann*

Acknowledgments

I am privileged to acknowledge Prof. James Day of Scripps Institution of Oceanography; Prof. Wilford Schmidt, University of Puerto Rico; and Dr. Kevin Fall for reviews, discussions, and suggestions. Steve Barilotti for introducing me to Patagonia with its diverse group of book editors and designers; author James Nestor's deep encouragements; Marianne Headley's decades of undying patience and support; Dr. Mark Stevenson's Mediterranean, transatlantic, and Pacific adventures; Bret Daniel's fluid wizardry; luminary Dr. Violeta Sanjuan for Tinetto exploits; watermen Brian Pucella and Scooter; Dr. John L. Newton's Arctic calmness; John Lockyer's airborne escapades; MiniTri enthusiasm of Dr. Griet Neukermans; Brock Rosenthal for years of conversations and hardware; Frederick DeLay for providing my first sail; Dan Beck's youthful explorations of Cannery Row; D.G. Wills for cultural sanity; CDR Thomas Keefer II for providing multicontinental adventure schedules; Palmer Station crew in Antarctica for keeping me safe; and for whatever "Big Chief" said in Dominica, "You have to make yourself happy and keep yourself alive."

Unfortunately, some of those who helped this rover navigate life are gone forever, departed, "over the bar": Bobbie Dean Wesson, surfing, sailing, windsurfing, and construction buddy; Evelyn Busuttil and her multicultural dynamics; Tiger (Lorilotte) Wiedmann Clark's intellectual guidance; and Prof. Walter Munk's decades of shared wisdom.

Willard Bascom celebrated the life of waves, I followed, encouraged by many others too numerous to mention here. Thank you all.

WILLARD BASCOM adventurer, inventor, engineer, and scientist—in the tropical Pacific during the Capricorn Expedition, 1952.
Roger Revelle Collection/UC San Diego Library

About the Author

Kim McCoy's ocean research began with surf zone wave dynamics, which led to pioneering advances in instrumentation, turbulence measurements, underwater communications, autonomous underwater vehicles, time reckoning, and freediving.

Educated in Germany, France, Britain, and the United States, Kim was presented with the Scientific Achievement Award (2018) as a Principal Scientist with NATO in Italy. Prior to Italy, Kim was CEO of Ocean Sensors, Inc. He interacted with the National Science Foundation, NOAA, NASA, WHOI, IFREMER, GEOMAR (and many other "acronym" institutions) and served as the Marine Technology Society's Chair for Ocean Instrumentation. Kim has planned projects around the globe, been awarded several patents, and is fluent in multiple languages.

Kim has examined beaches and observed waves on all seven continents; swam across the upper Amazon; journeyed along the Mekong, Nile, and Mississippi Deltas; traveled the Australian coastline; plunged into the Antarctic Ocean (without a wetsuit); crossed the Pacific, Atlantic, and Drake's Passage on ships; and sailed a boat from Africa to the Caribbean. He has been marooned twice (so far).

The adventure continues. Kim recently completed an Ironman and will continue to swim, dive, surf, sail, rock climb, and paraglide until motion stops, viscous drag ceases, buoyancy is lost, and gravity ultimately wins.

KIM MCCOY adventurer, oceanographer, engineer, inventor, sailor, freediver, paraglider, and polyglot, in Antarctica in 2002. *Kim McCoy Collection*

Index

A

ABS (acoustic backscatter sensor), 202
ACE (accumulated cyclone energy), 120–21
ADCP (acoustic Doppler current profiler), 200–201
Adriatic Sea, 112, 146
Aguaçadoura Wave Farm, 125
Agulhas Current, 82
Airborne Laser Bathymetry, 198
Airy, George, 61, 143
AIS (automatic information system), 183
Alaska, Gulf of, 79, 168
Aleutian Trench, 163
Alghero, Sardinia, Italy, 332
Amazon River, 145, 286, 310, 311
Amnisos, Crete, 161
Anaheim Bay, 295–96
Anak Krakatau, Indonesia, 173–74
Anchorage Cove, 168–69
Andrea Gail, 84
Antarctic Ocean, 82, 141
Arase Dam, 314
Arcachon, Bay of, 40
Arctic Ocean, 90, 124, 359, 376
Aswan High Dam, 308, 309
Atlantic City, New Jersey, 40
Atlantic coast (U.S.), 93, 140, 316
Atlantic Ocean, 25, 59, 114–15, 120, 133, 160, 173, 286, 310, 311, 322. *See also* North Atlantic; South Atlantic
Aysén River, 162

B

backrush, 266
backwash marks, 266, 267, 268
Badger, 168–69
Bagnold, R. A., 261
Baja California, Mexico, 34, 249
Baltic Sea, 146, 152, 342, 346
Banks Avenue, Mount Maunganui, New Zealand, 327
Banzai Pipeline, 216, 235, 237, 238, 390
barrier islands, 40–41, 42
bars, 42–43, 257, 258–66
Batang River, 145
bathymetry, 31, 106

Baton Rouge, Louisiana, 114
baymouth bars, 37, 38, 39
beaches. *See also individual beaches*
  anatomy of, 34
  berms and bars, 257, 258–66
  black-sand, 7, 35, 248
  definition of, 34
  dynamic nature of, 42
  extent of, 43–44
  famous, 273, 275
  formation of, 24, 33, 36–39
  geologic forms of, 36–43
  materials for, 248–52
  minor features of, 266–73
  nourishing, 304–5
  plastic debris on, 275–76
  pocket, 36, 37
  sea level changes and, 44–47
  white-sand, 34–35
beach face, 15, 43
Bedruthan Steps, Cornwall, United Kingdom, 134, 135
Beirut, Lebanon, 162
Bel Monte Dam, 310
Ben Cruachan, 83
Bengal, Bay of, 111–12
Benguela Current, 25
Bennington, USS, 76
Bensch, Chris, 372
Berge Istra, MS, 353
Berge Vanga, MS, 353
Bering Sea, 281
berms, 42–43, 45, 257, 258–66, 296–97
Bernoulli principle, 178, 179, 180, 196
Berouw, 172
Bhola (tropical cyclone), 112
Big Surf, 242
black-sand beaches, 7, 35, 248
Black's Beach, California, 189, 241, 243
BP, 358
breaking waves, 67, 76–77, 212–19, 327
breakwaters, 325, 327, 328–29, 331–35
Brunt-Väisälä frequency, 84–85
BSR Surf Ranch, 242
buoyancy, 62, 63

Burin Peninsula, Newfoundland, 162
Bush, George W., 113

C

cables, subsea, 345
Caldwell, Joseph, 288
California coast, 33, 36, 193, 225, 249, 288, 292, 293
Cannes, France, 249
Cantabria, Spain, 297
Cape Canaveral, 298
Cape Shirreff, Antarctica, 286
capillary waves, 27–28, 30, 66
Caravelas River, 298, 299
Carmel, California, 230, 259–61
Carmel River, 310–11
Carol, Hurricane, 113
Carteret Atoll, 122
Cenotaph Island, 168
Channel Islands, 293
Chelsea, Massachusetts, 153
Cherbourg, France, 325
Chesapeake Bay, 141, 280
Chicago, Illinois, 329
Chittagong, East Pakistan, 112
Chukchi Sea, 281
clapotis, 99–100
Clatsop Spit, Oregon, 298–99
climate change
    addressing, 122–23, 126–28, 372
    causes of, 47, 126–27
    effects of, 113–16, 120–25
    importance of, 21, 47, 116
Cloudbreak, 241
coasts
    classification of, 33
    definition of, 34
    development of, 307
    geomorphology of, 33
Cojo, 241
collapsing breakers, 217–18, 219
Colorado River, 314–15
Columbia River, 39, 229, 258, 324
Comfort, MOL, 351–52
continental shelf, 24, 25, 98
Cook, James, 233
Cook Inlet, Alaska, 301, 302
Coos Bay, Oregon, 259
coral reefs, 121–22, 123
Cornish, Vaughan, 78–79
Coromandel Peninsula, New Zealand, 4
Corson Inlet, 293
Cortés, Sea of, 314–15

Cortes Bank, 240
Costa, Steve, 300
Costa Concordia, 358
Cox, Charles "Chip," 88
Crescent City, California, 335
crest, definition of, 26
Cretan Star, 81
CTBTO (Comprehensive Nuclear-Test-Ban Treaty Organization), 166
Cudillero, Spain, 324
current ripples, 272–73
currents, 25
cusps, 109, 266, 269–71

D

dams, 305–11, 313–14, 318
Danel, Pierre, 333
Decatur, Alabama, 242
Deepwater Horizon, 358–59
deep-water waves
    definition of, 57
    energy in, 262, 326–27
    group velocity of, 94–95
    measuring, 195, 202
    shallow-water vs., 57, 99
    swell, 92–95
Delaware, USS, 180
deltas, 305, 314–19
Derbyshire, MV, 355
Destin, Florida, 35
Dexter, Phil, 242
Dhu Heartach Lighthouse, 323
diffraction, 98, 101–2
dikes, 327
doldrums, 25
domes, 266, 271
Dorian, Hurricane, 115
Dorset, United Kingdom, 109
Dover, Strait of, 303
Drake Passage, Southern Ocean, 52, 66, 68–69
drowned topography, 33
Duck, North Carolina, 208
DUKWs, 184–85, 202, 205, 225–32
dune ridges, 298–99
dunes, 41–42, 43, 44, 302–3, 334–37

E

Earth
    albedo of, 121
    formation of, 21, 23
    geology of, 23–24, 32–33
    orbital movement of, 45
    rotation of, 24–25

tilt of, 45, 139, 140
    water on, 23, 24
earthquakes, 20, 159, 166–68
East River, 154
Eckart, Carl, 108
Eckman, V. W., 183–84
Eday Island, 154
Edmund Fitzgerald, SS, 355
El Faro, SS, 354
El Niño–Southern Oscillation (ENSO),
    26, 290
Elwha Dam, 313
Encitas, California, 285
English Channel, 154
Eniwetok Atoll, Marshall Islands, 174
Ephesus, Turkey, 46
Esso Languedoc, 80
explosion-generated waves, 171–75, 176–77
extreme waves. *See* rogue waves
Exxon Valdez, 357

F
fetch, 67, 68
Fincham, Adam, 243
FLIP (Floating Laboratory Instrument
    Platform), 188, 349–50
Florence, Hurricane, 122
Foo, Mark, 241
Fort Bragg, 249
Fort Ord, 259–60, 262, 270, 335
Franklin, Benjamin, 185
Freeth, George, 233
Fukushima Nuclear Power Plant, 162,
    170–71, 174
Fukushima Renewable Energy Institute, 171
Fundy, Bay of, 140, 145

G
Gaillard, D. D., 323, 325
Galileo, 133
Galveston, Texas, 40, 110
Ganges-Brahmaputra Delta, 112, 314,
    317, 319
Gaoping River, 313
Gary, Indiana, 325
Gerstner, Franz von, 53–54, 61
Gibraltar, Strait of, 219
Glines Canyon Dam, 313
Golden Gate, 141, 154, 302
Goose Island, 280
Grand Banks of Newfoundland, 83, 162
Grand Ethiopian Renaissance Dam, 308
Grand Harbour, Valletta, Malta, 384
gravity, 27, 62, 63, 133

Great Exuma, Bahamas, 275
Great Pacific Garbage Patch, 276–77
Green Harbor, Massachusetts, 328
groins, 303–5
group velocity, 94–95
Guiuan, Philippines, 116
Gulf Coast (U.S.), 40, 122
Gulf Stream, 25, 83, 142
gyres, 25

H
Hagibis, Typhoon, 120, 171
Haiyan, Super Typhoon, 115, 116
Half Moon Bay, California, 240, 241, 263
Hall brothers, 183
Hamilton, Laird, 239
Harvey, Hurricane, 114, 313
Hatteras Island, North Carolina, 199
Healy, USCGC, 376
Hendricks, Terry, 235
Henwood, J. W., 282–83
Hilo, Hawai'i, 164, 165, 166, 167
Hintze Ribeiro Bridge, 313
Höfn, Iceland, 200
Holthuijsen, Leo, 81
Honolulu, Hawai'i, 346
Hudson River, 33, 185
Humboldt Bay, 226, 227, 337–40
Humboldt Current, 25
hurricanes, 113–15, 369. *See also individual
    hurricanes*
hydrocarbon production, 343, 345–46,
    358–59
hydrodynamics, 53

I
ice, 89–90, 92, 123–24
IHI Corporation, 154
IMO (International Maritime Organization),
    360
Indian Ocean, 25, 140, 141, 159, 162, 170,
    173, 317, 359
Indus River, 314
Inman, Douglas, 292
internal waves, 84–87, 182–84, 209
Iquique, Bay of, 162
Irma, Hurricane, 114
irrotational theory, 61
Isaacs, John, 108, 125, 126, 202, 225–27,
    232, 301
"Ivy Mike" thermonuclear explosion, 174,
    176–77

J
Jakarta, Indonesia, 317

James Island, 298
Jaws, 240
Jebi, Typhoon, 128
jetties, 327, 328, 330, 337–40
John, J. W., 256
Jones, William H. S., 80
Juan de Fuca, Strait of, 313, 337
Junger, Sebastian, 84

K
Kahanamoku, Duke, 233, 234
Kalama, Dave, 239
Kamilo Beach, Hawai'i, 275, 276
Kansai International Airport, 128
Katrina, Hurricane, 113–14, 342
Kaua'i, Hawai'i, 388
Kelly Slater Wave Company, 243
kelp, 88, 89
Kelvin, Lord, 177, 178, 192–93
Kelvin waves, 178
Kepler, Johannes, 133
Kiribati, 121–22
Kirra, Gold Coast, Australia, 9
Kitty Hawk, North Carolina, 41
Klamath River, 39, 314
Klein, H. Arthur, 233
Koxa, Rodrigo, 239
Krakatau, Indonesia, 162, 171–73
Kuroshio Current, 25

L
Laguna Beach, California, 364
La Jolla, California, 124, 241, 252, 253, 365
Land's End, 322
landslides, 159–60, 161
La Rance Bay, 153, 154
lasers, 198–99
Leadbetter Beach, California, 288, 290–91
Leadbetter Spit, Washington, 258–59, 265
Lemoore, California, 243
Leo Carrillo, California, 88
Leviathan, 180
Lidar, 198, 199
lighthouses, 323–25
Liquid Robotics, 209
Lituya Bay, 168
Long Island coast, 38, 281, 304
Lopez, Gerry, 237
Los Angeles, California, 152, 167, 240, 241, 288, 294
Lothal, India, 110
Louis S. St-Laurent, 376
Lunada Bay, 240
Lyell, Charles, 283

M
MacInnis, Joe, 355
Mackenzie River, 314
Maeslantkering, 340, 342
Makassar, Celebes, Indonesia, 162
Malibu, California, 241
Mallory, J. K., 82
Malloy, Keith, 368
Man, Isle of, 281
Maria, Hurricane, 111, 114
Massachusetts Bay, 87
mass transport, 61
Mavericks, 240, 241
McNamara, Garrett, 239
measurement, units of, 15–16
Mediterranean Sea, 45, 56, 133, 149, 162, 166, 174, 219, 358, 359
Mekong River Delta, 315–16
Mentawai Islands, Sumatra, Indonesia, 208
Mexico, Gulf of, 110, 113, 204, 316, 343, 358–59
Miami Beach, Florida, 40, 273, 336–37
Michigan, Lake, 325, 329
Mihama, Mie Prefecture, Japan, 120
Milankovitch cycles, 26, 45
Milford Sound, New Zealand, 50
Mississippi River, 314, 316–17
Montauk Point Lighthouse, 281
Monterey Bay, 232, 262
Monterey Harbor, 152, 153, 163–64, 262
Moon
    gravitational pull of, 20, 21, 133–38
    orbit of, 135, 137, 138, 139
    rising of, 139
Morro Bay, California, 335
Mull, Isle of, 132
Munch-Petersen, J., 287
Munk, Walter, 69, 108, 124–25, 193, 202

N
Namib-Naukluft National Park, Namibia, 44
Nansen, Fridtjof, 20
Nantucket Island, 140
Napoli, MSC, 352
Narragansett Bay, 113
National Oceanic and Atmospheric Administration, 120, 165, 288
natural gas production, 343
Nazaré, Portugal, 239, 240
Nearshore Sediment Transport Study, 288–89
Neptune Sapphire, 83
Neumayer Channel, 69

Neva River, 146
New Haven, East Sussex, United Kingdom, 71
New Orleans, Louisiana, 113–14
Newport Beach, California, 241
Newton, Isaac, 133
New York Harbor, 38, 40, 41
Niger River, 310
Nile River, 308, 314
North Atlantic, 45, 71, 78–79, 239, 240, 358
North Brazil Current, 310
North Pacific, 238, 240, 277, 337
North Sea, 110, 204, 302, 303, 327, 340,
    342, 343, 352, 359
Noyo River, 152
nuclear explosions, 174, 176–77
Nusa Penida, Bali, 37

O
O'Brien, M. P., 225, 330
OBS (optical backscatter sensor), 198–99
The Ocean Cleanup, 276–77
Ocean Motion Technologies, 126
The Ocean Race, 357
Ocean Ranger, 358
offshore structures, 343, 345–46
Offshore Technology Research Center, 208
oil
    effects of, on waves, 87–88
    production of, 343, 358–59
    spills, 357, 358–60
Omaha Beach, 258
Orange River, 44
Orchid, Typhoon, 171, 355
Oregon coast, 36, 193, 218, 225, 249,
    261, 372
orthogonals, 107
oscillation ripples, 273
oscillatory waves, 53

P
Pacifica, California, 122, 285, 286
Pacific Basin, 115, 160, 163
Pacific coast (U.S.), 10, 39, 93, 106, 122, 143,
    163, 233, 258, 316, 330, 338, 365
Pacific Grove, California, 163–64
Pacific Ocean, 4, 25, 35, 86, 115, 140, 141,
    151, 160, 167, 170–71, 174, 176, 193–94, 217,
    226, 234, 290, 302, 314, 337, 338, 354,
    359. See also North Pacific; South Pacific
Pago Pago, Samoa, 151
Pam, Cyclone, 4
Panama Canal, 140–41
pancake ice, 90
Pardo, Arvid, 127
Pausanias, 161

Perry, Robert, 90
Persian Gulf, 81, 359
Petrobras-36, 358
Phuket, Thailand, 170
Pieman Heads, Takayna/Tarkine,
    Tasmania, 20
pinholes, 266, 270, 271
Pipeline. See Banzai Pipeline
pipelines, subsea, 345–46
Pisa, Italy, 46
plankton, bioluminescent, 364
plastic debris, 275–76
plate tectonics, 23
Plenty, Bay of, 351
Plimsoll, Samuel, 350
Plimsoll mark, 349, 350
plunging breakers, 215–16, 235
pocket beaches, 36, 37
Point Conception, 240, 241, 249, 288
Point Mallard Park, 242
Point Mugu, 288, 298
Polynesians, ancient, 50, 384
Port Hueneme, California, 293, 294
Porto da Baleeira, Sagres, Portugal, 333
post-glacial rebound, 32, 46–47
Prince William Sound, 357
Providence, Rhode Island, 113
Puget Sound, 33
PWCs (personal water crafts), 231, 239

Q
Qiantang River, 144–45
Queen Mary, 81
Quillayute River, 298

R
races, 355, 357
radar, 199–200
Raglan Township, New Zealand, 28
Rankine, W. J. M., 61
Raubenheimer, Britt, 189
Redondo Beach, California, 233
reflection, 98, 99–101
refraction, 98–99, 103–7, 213, 296–98
Rena, MV, 351
Revelle, Roger, 202
ribbons, 303
rill marks, 266, 267, 268, 269
Rincon, 240
rip currents, 221–25, 229
ripple marks, 266, 271–73, 275
ripples, 26, 27, 66, 272–73
rivers, 305–6, 309–11, 313–14.
    See also individual rivers

Rockaway, 38, 40, 41
rogue waves, 79–84
rotational flow, 61
Rotterdam, Netherlands, 327, 340
Russell, John Scott, 55–56, 95, 185, 218

S
Sahara Desert, 310
Saint Petersburg, Russia, 146, 342
saltation, 63, 300
Salt House Beach, Norfolk,
  United Kingdom, 45
Sanchi, 355
Sanchis, Benjamin, 239
San Clemente Dam, 310–11
sand
    beach slope and, 262, 263
    creation of, 282–83
    definition of, 35–36, 249
    forms, curious, 296–99
    grain sizes of, 250, 251
    littoral drift of, 292–96
    longshore transport of, 286–92
    offshore-onshore motion of, 252–57
    in tidal entrances, 299–303
San Diego, California, 122, 124, 167, 170,
  240, 305
Sandon Point, New South Wales,
  Australia, 103
Sandy, Hurricane, 41, 114
Sandy Hook, 38, 40, 41
San Francisco Bay, 141
San Onofre, 241
San Simeon, California, 270
Santa Barbara, California, 101, 152,
  288–91, 293
Santa Clara River, 295
Santa Cruz Island, 104
Santa Maria River, 306
Santa Monica, California, 291, 294
Santorini, Greece, 162, 174, 178
Schettino, Francesco, 358
Scotch Cap Lighthouse, 165, 325
Scotrenewables Tidal Power SR2000, 154
Scripps Institution of Oceanography, 69,
  75, 86, 108, 126, 207, 222, 292, 300,
  349–50
Scripps Ocean Atmosphere Research
  Simulator, 207
sea level
    changes in, 44–47, 113, 121–25, 127, 281
    global, 143
    local, 142
    mean, 143

seawalls, 327, 332, 334, 336
sea waves, 69, 71–77. *See also* seismic
  sea waves
seiches, 146–53
Seine, 145
seismic sea waves, 159–65
Severn, River, 143–44
Seymour Narrows, British Columbia,
  Canada, 141, 142
shallow-water waves
    deep-water vs., 57, 99
    definition of, 57, 98
    diffraction of, 98, 101–2
    energy in, 107, 326–27
    measuring, 195
    reflection of, 98, 99–101
    refraction of, 98–99, 103–7
    storm surges, 109–13
    trapped waves, 107–9
    shear, 63
Shepard, F. P., 222
Shetland Islands, 325
ships
    disappearance of, 354–55, 360–61
    metacentric height of, 348
    motions of, 347–48
    Plimsoll mark for, 349, 350
    sinking, 350–54, 355, 357–58, 360
    stability of, 348–49
    steadiness of, 348
    waves produced by, 177–85
shoreline
    definition of, 34
    erosion of, 280–86, 292, 295–96, 303–5
shoreline structures
    design of, 327–37
    effects of waves on, 325–27
Sihwa Lake Tidal Power Station, 153
Slade's Mill, 153
Slater, Kelly, 243
Slocum Thermal Glider, 209
Smugglers Cove, 104–5
Solana Beach, California, 285, 305
solitary waves, 218–19
South Atlantic, 141
Southern Ocean, 52, 66, 68–69, 79, 357
South Pacific, 141
spilling breakers, 216–17
spits, 38, 39
stars, navigation by, 59
Steamer Lane, 241
Stellwagen Bank, 87

steps, 266, 269

Stockholm, Sweden, 346

Stokes, C. C., 61

Stokes drift, 61

storm surges, 109–13, 116

storm waves, 77–79

Styx River, 145

submarines, 180, 181–82

Sumatra, Indonesia, 159, 162, 168, 169–70, 172, 212

Sun
  gravitational pull of, 20, 21, 133–34, 137, 138–39
  radiation of, 24, 45

Sundra Strait, 172–73

Sunmore, 168–69

surface tension, 27–28, 88

surf beat, 219–21

surfing
  artificial, 242–43
  best areas for, 235, 237, 238, 240–42
  big-wave, 238–41
  in DUKWs, 229–32
  films, 238
  history of, 233–34
  objective of, 234–35
  popularity of, 234, 237–38, 244
  on ship-produced waves, 185
  strapped (tow-in), 239

Surf Snowdonia, 242

surf zone, 15, 212, 233

surging breakers, 215

Sverdrup, Harald, 69, 202

swash marks, 266, 267, 268

swash zone, 43

swell, 92–95, 212

T

Table Bluff, 226, 232, 259, 270

Taiohai, Marquesas Islands, 164

Taylor, David, 180

Teahupoʻo, Tahiti, 219, 368

Tempe, Arizona, 242

tetrapods, 333, 334, 335

Te Whanganui-A-Hei Marine Reserve, 4

Thiermann, Kyle, 241

Thorogood, John, 360

Three Gorges Dam, 309

tidal bores, 143–45

tidal currents, 141–42

tidal entrances, 299–303

tidal prism, 330–31

tide gauges, 192–93

tides
  cause of, 21, 133–40
  coastal configuration and, 140–41
  definition of, 132
  harnessing power of, 153–54
  high, 133, 134, 138
  king, 139
  low, 133, 135
  neap, 137, 138, 139
  period of, 26, 133
  range of, 133, 138–41
  spring, 137, 138, 139

Tillamook Rock, 323–25

Tinetto, 149–50

Tōhoku tsunami, 158, 162, 170–71

tombolos, 297–98

Toronto, Ontario, 346

Toshiba Corporation, 154

trapped waves, 107–9

trochoidal waves, 54

trough, definition of, 26

tsunamis
  causes of, 20–21, 158, 159–60, 161, 171
  etymology of, 21
  historic, 161, 162–65, 167–74
  period of, 26
  seiching and, 151
  terminology for, 158–59, 163
  velocity of, 160
  warning systems for, 165–71

Tubbataha Reef Natural Park, Philippines, 123

Twin Rocks, Oregon, 220

typhoons, 115

U

UNCLOS (United Nations Convention on the Law of the Sea), 127, 276, 360

undertow, 221–25

Unimak Island, Aleutian Islands, Alaska, 165

uprush, 266

US Army Corps of Engineers, 205, 288, 289, 291, 337–38

US Coast and Geodetic Survey, 162, 165, 192

V

variables, 16–17

Vendée Globe race, 355, 357

Venice, Italy, 112, 113, 146

Ventura Harbor, California, 293–95

viscosity, 62, 63

volcanoes, 171–74

W

Waco, Texas, 242

Wafra, 83

Wai'ānapanapa State Park, Hawai'i, 248

Waikiki Beach, Hawai'i, 216–17, 233, 234, 273, 304

Waratah, SS, 82

Washington, George, 281

Washington coast, 193, 218, 225, 249, 298

Wateree, USS, 162

wave channels, 55–56

Wave Glider, 209

wave height, 26, 56, 58

wave length, 26, 56–57, 58

wave makers, 205–8, 242–43

wave period, 26–27, 28, 56–57, 58

waves. *See also* deep-water waves; shallow-water waves; tides; tsunamis; wind waves

breaking, 67, 76–77, 212–19, 327

buoyancy and, 62, 63

capillary, 27–28, 30, 66

causes of, 20, 66

complexity of, 30–32, 50–51

concept of, 20

explosion-generated, 171–75, 176–77

force of, 202–5, 326–27

forecasting, 69, 75

fundamental properties of, 56–59

gravity and, 27, 62, 63

harnessing power of, 125–26

ice's effects on, 89–90, 92

internal, 84–87, 182–84, 209

mass transport and, 61

measuring, 191–205

observing, 189–92

oil's effects on, 87–88

orbital motion and, 59–61

oscillatory, 53

parts of, 26, 27

research on, 51–56, 61, 69, 86, 188–92, 368

rogue, 79–84

in a sea, 69, 71–77

ship-produced, 177–85

sizes of, 25, 26–27, 77–79

solitary, 218–19

storm, 77–79

of translation, 55–56, 218–19

trapped, 107–9

vehicles powered by, 209

velocity of, 57

viscosity and, 62, 63

wave sets, 219

Waves Project, 190, 193, 225, 258

wave staffs, 192, 196

wave steepness, 58, 67

Weber, Ernst and Wilhelm, 54–55

Weddell Sea, Antarctica, 90

The Wedge, 241, 242

White Cliffs of Dover, 32

white-sand beaches, 34–35

Wight, Isle of, 123

Wigley, W. C. S., 180–81

windfarms, 343, 345

winds

cause of, 25

directions of, 25

measuring, 369

wind waves

creation of, 66–67

internal waves, 84–87

rogue waves, 79–84

sea waves, 69, 71–77

size of, 67–68, 77–79

storm waves, 77–79

swell, 92–95

variability of, 20

Wirewalker, 209

Woodbridge Tide Mill, 153

Woods Hole Oceanographic Institution, 86, 171

World Glory, 83

World Horizon, 83

Y

Yabucoa Harbor, 114

Yangtze River, 309, 314

Yellow River, 314

Yolanda, Super Typhoon, 115

Z

Zoe, MSC, 352

NEXT SPREAD: No matter how much we resist, the power of the waves always prevails. *Bob Barbour/Nat Geo Image Collection*